THE CHRISTIAN'S Y2K PREPAREDNESS HANDBOOK

ISBN 0-967-0136-0-7

Published in the United States by
The Home Computer Market, Inc.
PO Box 385377, Bloomington, MN 55438.
www.homecomputermarket.com

We dedicate this book to our children,
Anna (who came up with the idea),
Bethany, Charity, David, Elijah, Philip,
and Geremiah, who waited patiently
(and sometimes not so patiently) for
mommy and daddy to finish "the book."

If you would like additional copies of this book,
or more information about Y2K,
check out our web site at:
www.homecomputermarket.com
or write or call us at
Home Computer Market
PO Box 385377
Bloomington, MN 55438
612-844-0462

We are also available to do Y2K Preparedness Workshops.

THE CHRISTIAN'S Y2K PREPAREDNESS HANDBOOK

Section III
What To Do

Section IV
Your Money and Y2K

Section V
Your Church and The Church

Section VI

THE CHRISTIAN'S Y2K PREPAREDNESS HANDBOOK

Introduction

The first time my wife brought up the year 2000 computer problem, I told her it was no big deal, it was just something a little re-programming of software would fix. She showed me an article in a Christian newsletter that almost implied the world as we know it would come to an end ("The-end-of-the-world-as-we-know-it" is commonly referred to as the acronym TEOTWAWKI) as a result of this little computer glitch. "This is ridiculous! Pure hysteria!" I told her. "Computer programmers have know about this for years, and they are already fixing any problems. We don't need to worry about it. That article is just another one of those wacko, paranoid, the end-is-near theories." Well, after much research into the issue, it turns out both the article and myself were wrong. The truth lay somewhere in the middle.

Not long after my wife brought up the year 2000 problem we started getting calls about it from people we already knew. They were calling me because they knew that for many years, before we started the Home Computer Market, I was a full-time rare coin and gold and silver dealer. These people were calling me to get counsel on buying gold and silver before the year 2000. They were all convinced we were facing certain economic collapse, along with many other disasters. They believed gold and silver would be the only way business would be able to be conducted after January 1, 2000 because the dollar would be worthless (or at best severely devalued) due to year 2000 problems. Hence, they wanted me to give them advice on buying gold and silver coins for barter.

Which all brings me around to answering the question "So why write a book?" The question is especially relevant since there are already books, web sites, articles, and newsletters on the subject of the year 2000 problem. Someone, who had been overhearing me giving all these people counsel said, "You should write a book." And they were right. We felt compelled to write a book on Y2K because we felt what needed to be said in the Christian community was so different than what is generally being published in most other sources. And we found much of the preparation advice being given was what we believed was actually either bad advice, or not what we believed God would have us as Christians do. As we were doing our research, we found surprising little material approaching this from a Christ-centered viewpoint. Most of it approaches the subject from one of three perspectives:

- Technical explanation of the causes and effects of the year 2000 computer glitch; including varied speculations as to what will happen January 1.

- Technical discussion of solutions and fixes to various year 2000 related problems.

- Sell everything, get your gold and silver, get your year supply of food, get your guns, and get to the hills! The end is near!

Most information that we were able to find in Christian circles on this issue seemed to have some kind of spin on the third perspective. We really believe most advice we are hearing, and being suggested from most Christian sources, is not the best, or even wise. We do not believe the "sell all, buy gold, silver, food, guns and head for the hills" (or the woods, or the prairies, or whatever isolated rural area) advice is correct. And it is probably not even Scriptural. God calls us to be in the world, but not of the world. We are to be a witness, not an isolationist. Yes, there are computer experts who are of a "get out of town, civilization as we know it will end January 1, 2000" type of mindset. But *most* computer experts do *not* hold such an "Armageddon" opinion of what will result of all this.

We believe the most important issues related to the Year 2000 problem are not things such as buying gold & silver or how much food will we need and how we should store it. Although these things are important to deal with, there are other things, which have far greater importance in the light of eternity.

We will cover the above issues and many others. We set out to make this a practical preparedness Handbook to assist Christians in planning for both the least and the worst of what may come out of all this. We have five ultimate goals we are trying to help Christians achieve:

- Knowing we are truly are a Christian and will spend eternity in Heaven.

- Getting involved with the best local church possible.

- The awareness and concern of the Year 2000 issue among the general population translates into the greatest witnessing opportunity of our generation. We believe that how we respond (both individually and as a church) to this issue will have a tremendous effect on our witness to a lost and unsaved world. This Handbook will show how to be the best witness to an unsaved world through this situation.

- The terrible risk we all run in destroying any possibility to witness through our potential responses to the whole Y2K issue. The concern Tammy and I have is that many Christians who do start altering their lives will be wrongly labeled as "religious wackos" or "kooks" by many of those around them. We plan on addressing this issue (which is not being addressed in most of the Christian material we are reading on the subject), the dangers of being labeled a "religious wacko," and how to avoid getting that label. We believe this is also one of the most important issues of our book.

- How to be best prepared and protect our families and

investments whatever may come, without significant expense or trouble over and beyond what you would spend anyway and how to prepare, from a Christian perspective, for the rest of your life in the 21st century.

My wife and I admit we are not computer programmers or computer geniuses. In fact, many of you who read this Handbook will possess greater computer knowledge and expertise then ours. If our technical analysis is not deep enough for you, there are other books out there from people who are better qualified to go into the academic explanations. It is not our goal to convince you of this problem from a technical viewpoint. But we have done our research into the issue of the year 2000 problem. We will do our best to explain the problem, its causes, its solutions, and what you should do to prepare. All in plain English. And, many of the concepts and suggestions we put forth in this Handbook are not meant just to help you through January 2000. They are meant to give you direction for your entire journey through the next century.

With all that said, we also believe everyone needs to have a contingency plan just in case the worse case does happen. What will you do? You better know ahead of time.

One thing we cannot stress enough is that *everyone* who has investigated the issue, from government experts to business experts and to computer experts, all agree on this one thing: it is not a question if Y2K will effect you, the question is how it will effect you. The two questions we ask you are "What are you going to do to prepare?" and "As a Christian, what does our Lord Jesus Christ want you to do?" We trust this Handbook will give you the answers that will best bring you and your family through this problem, the best way possible.

Your servants in the Lord,
Dan and Tammy Kihlstadius

SECTION I

THE PROBLEM

CHAPTER 1

WELCOME TO Y2K
OR
READY OR NOT,
HERE IT COMES!

WHAT IS Y2K?

What exactly is this problem that computers will have when the clock turns to 12:00 A.M. January 1, 2000, and why is it referred to as Y2K?

$$Y = Year$$
$$2 = 2$$
$$K = 1,000$$
$$2K = 2 \times 1,000$$
$$Y2K = Year\ 2 \times 1,000$$
$$Y2K = Year\ 2000$$

The "Y" represents the word "year." 2K is a commonly used abbreviated term for the number 2,000. Hence Y2K is an abbreviated form of the year 2000.

What does Y2K have to do with computers?

Long ago, when computers first became a part of our business and government infrastructures, they were extremely expensive. Every

3

part was expensive, programming was expensive and building them was expensive. Computing power in 1960 was infinitesimal compared to today's supercomputers. Whereas now a common desktop computer will set you back about $2000, a computer of comparable power was unimaginable in the 60's. The fastest and biggest computers back then would cost millions and occupy an entire city building. To work with such limited computing power, programmers had to be extremely clever and efficient at creating programming code to serve management's purposes. Computers have essentially 4 parts; an input/output, hard drive (or long term data storage), a processor (the brain) and RAM (Random Access Memory or the temporary data storage). All of these parts were very limited in size and "bigger" ones were either nonexistent or not in the budget. "Computer storage cost ten thousand times (that's 1 million percent) more than it does today." [1]

Many of the computer's functions involved handling dates. The date of a transaction, the date of birth, the future date of a planned action, the past dates of previously occurring actions, and so on. All these dates needed to be stored in the computer and remain accessible for the programs to function. To maximize storage space, most if not all, computer programmers, stored and used information about dates in this format: mmddyy. Whereas we read a date like March 21, 1961 as such, a computer would read it as all numbers: 032161. Only two "spaces" in the date "field" were allowed for the year because we all knew it was the 1900's. This two-digit year format became a standard and is now found in many programs and computer systems.

Critics of this programming method argue that programmers were short–sighted and incompetent in choosing this method of storing dates. However, remember many of these programs were written in the 50's, 60's, and 70's. Called "legacy" programs, they remained in service for years, updated and added onto as the system requirements and application goals changed over the years. In our opinion, the fact that these old programs are still used is a testimony to the original programmer's skill and ability. They are still the programs that banks, companies, government agencies, and other have in service. Not bad for programming code written over 30 years ago!

However, now that the year will change over to 2000, we face many computer problems. Why is this a problem you say? Because many of these programs or applications use dates for time sensitive calculations, this change of year (1999 to 2000) will cause problems. Under normal circumstances, every year the program would automatically increment the year number by 1 and thus, 99 will have a 1 added to it. However, most people visualize this date sequencing to be like an odometer, slowly turning to 00. In reality, the date *space* in most programs can only handle 2 digits and there is no 2 digit number higher than 99. Thus, some of these programs are actually very likely to fail or give erroneous results, rather than perceive the date to be 1900. Think of like a full glass of water. The two digit field will be "full" at "99." If you add anything to it, just like adding to the full glass of water, you will have a mess.

This is how one Y2K computer expert explained it:

"Without directly quoting, just let me say that yet another major publication has stated that computers will get confused at the witching hour of 1/1/00 and be unable to distinguish between 1900 and 2000. At this stage I'd venture a guess that it seems to be a knee jerk cut and paste activity for Y2K writers to use the standard phrase: 'computers will be unable to distinguish 1900 from 2000.' Unfortunately this is not what will happen. For one thing, computers really do not get 'confused.' That's placing a degree of anthropomorphism on inert machinery that credits them with a degree of intelligence that simply does not exist. If computers were smart, they'd trip over bogus date math and gently raise their hands: 'Excuse me sir! Do you really mean this?' Not very likely. The problem with software processing "OO" dates (and beyond) is the long-standing convention of defining dates as some variation of the yymmdd form. The year field-yy-is a two-digit number. Given that we have explicitly stated that we are working with two digits only, this means we have a potential mathematical series running from 00, 01, 02... through ...97, 98, 99. Full stop!!!" says David Eddy.[2]

Another fallacy concerning the Y2K problem is referring to it as the "millennium bug." It is not a "bug." A bug is a computer term used to describe an unexpected failure or glitch in a computer program usually due to an unintended human error in the programming of the software. On the other hand, the Y2K computer problem was a very methodical programming tactic designed for a specific purpose: to save time and money. And it did just that. However, no one anticipated that these systems would still be in use as much as 30-40 years later and now we must convert all the date fields to a four-digit format. This is the crisis: too many programming lines to fix and not enough time.

This two-digit format can be found in multitudes of computer systems and microchips, in some way affecting many governments, utilities and industries in the civilized world. Software engineer Capers Jones estimates that about 25% of US software applications are assumed to contain year 2000 problems.[3] Computer programmers and scientists are working feverishly to correct these systems before January 1, 2000. Jones estimates that only 85% will be repaired in time. Therefore, while most Y2K doomsdayers want you to believe Y2K problems permeate the entire computer infrastructure, this computer expert expects that less than 4% of all software applications will suffer Y2K problems at midnight on December 31, 1999.[4] We will discuss specific industries, utilities and governments, how the Y2K problem affects them, and their failure potential in a later chapter.

But how much of Y2K is... Hype, Hysteria, or Fact?

First, a little background on one of the authors, Dan.

Over the years, as a dealer in gold, silver, and rare coins, Dan has dealt with many different kinds of people. And he learned early on that some of those people bought gold and silver coins (and only coins, not bars or ingots) for some strange reasons. Many of these strange reasons revolved around two basic premises. #1) The value of the U. S. Dollar was about to collapse and/or we were facing imminent hyper-inflation in just a few months (like in Germany in the 1920's) So gold or silver would be the new form of money and used as a means of a bartering system. Or, #2) much of the entire economy (both U.S. and worldwide) is under the control of some super secret organization and the only way to protect yourself financially was with gold or silver coins.

After a while, Dan had heard enough of this. He realized there was never any real evidence to these claims. Usually these theories came from people with little or no credibility. It was an article written by a part time coin dealer that made Dan lose total trust in gold and silver as a replacement for Federal Reserve Notes, or dollars, or cash (as we will be referring to those green pieces of paper you use to buy things at the store.) Reading this one article of logic and reason was what really opened up his eyes to the error of thinking gold and silver would be the best means of "monetary" exchange.

So are we going to tell you what the article said which convinced us that almost everyone buying gold and silver coins for the previously mentioned reasons were all wrong? Yes, and no. Yes, we will tell you, but not now. More about gold and silver will come in Chapters 12 and 13 in the investment section. You will have to wait.

The reason we are bringing this all up now is to give you a small picture of some of what we saw going on around us for many years. The cry of "Sell all, get your gold, and get to the hills!" is nothing new to us. We first started hearing doomsday hype in 1979. We also bring this up to let you know that at one time or another, we have heard some form of almost every doomsday, economic collapse theory. Of every theory we've ever heard, Y2K is the only one that we are listening to, learning about, and doing something because of it.

Hype?

Over the years, many prophets of doom have cried out, "The end is near! Prepare!" With repeated warnings, the listener begins to doubt the validity of such claims. In the 80's, Gary North predicted the ultimate collapse of our economy. During the recession of 1973-1974, pessimistic prophets predicted we would run out of fossil fuels within 10 years. These prophets urged all of us to turn our thermostats back, drive 55 and buy subcompact cars. Before that, others have predicted global nuclear war was unavoidable so people were exhorted to build bomb shelters in their backyards. In the forties, many Christians speculated that Adolph Hitler was the anti-Christ and he was on his way to securing leadership over a one-world government. One hundred years ago, reputable scientists predicted massive famines within years if the population rose over one billion. None of these doomsday scenarios ever materialized. The result is that our ears have become deaf to these exhortations. The story of the boy who cried wolf is very accurate in its portrayal of human nature. There is a lot of hype in the multitude of Y2K publications and Internet sites. However, the Y2K crisis is not one to be ignored. It is real.

However, because of the numerous sensationalized media reports, many of the average Joes on the street have already turned a deaf year. Hearing hyped–up reports of planes falling out of the sky, elevators crashing to the ground floor, nuclear reactors exploding, power outages causing electrical appliances to blow up, etc, just numbs most people's minds. In fact, these types of overstated disasters do nothing to promote Y2K awareness or preparedness. One Y2K expert put it this way: "Although their creators are undoubtedly well intentioned, it is important to point out that supposedly helpful analogies for explaining Y2K risks can backfire in a major way. What comes to mind is the oft-repeated and now cast-in-stone as an Urban Myth that elevators will cease working at the 1/1/00 witching hour. Even worse, they will crash to the basement level. Personally, I think the man in the street hears this dire prediction and is just further convinced that Y2K is a total hoax." This Y2K expert's conclusion (remember, these are the expert's own words): "Consequently, EveryMan—-from JoeSixPack to CharlesCEO—-heavily discounts any urgency for Y2K action and goes back to watching The X-Files."[5] We need to explain Y2K in a rational, calm way using facts and verifiable information, not overblown hype. Some people only hear the negative hype and never any of the positive progress that is being made on Y2K conversion.

Hysteria?

On the other hand, there seems to be a great deal of hysteria and panic in some Christian circles. We have read many Christian and non-Christian resources regarding the Year 2000. In most Christian sources of information on Y2K, several over-riding messages ring all too clear.

Let us point out the weaknesses and errors in these messages:

1. *"The year 2000 computer glitch is God's judgment on our computer dependent, technology worshipping society."* We feel it is presumptuous of us to assume we know *why* this Y2K event is occurring.

It is an amoral event. It has no relation to any sin whatsoever. Shortsightedness maybe, but not any blatant immorality. (AIDS could be called a judgment of God on homosexuals. But a programming mistake is not a judgment.) Back in the early days of computing, the people who hired programmers forced the programmers to create the most memory efficient software. "The year 2000 problem actually originated as an explicit requirement by clients of custom software applications and the executives responsible for the data centers as a proved and seemingly effective way of saving money."[6]

Some prominent programmers (including the man who invented the programming language of COBOL) have revealed that they pointed out the potentially dangerous consequences of a two-digit date field as early as 1969! No one listened because there was not enough memory or money to rectify the problem at that time. In addition, no one could foresee the way computers have been integrated into the machinery of our global society. We have seen studies that show that the amount of money saved by using a 2-digit format for the last 40 years is almost equal to the current costs of fixing it! Capers Jones points out, "Dr. Leon Kappelman has calculated that the accrued savings of the past 30 years and the current year 2000 repair costs are roughly commensurate."[7] So, who's to say it wasn't a smart thing to do? We just waited too long to fix it!

To say the Y2K crisis is God's judgment is to put the face of sin on a neutral programming quirk. In Luke 13:4-5 Jesus explains that bad things do not happen to people just because they are evil. "Or those eighteen, upon whom the tower in Siloam fell, and slew them, think ye that they were sinners above all men that dwelt in Jerusalem? I tell you, Nay: but, except ye repent, ye shall all likewise perish." Things happen to people and these things sometimes happen independently of their morality. As it is said, God makes it to rain upon both the just and the unjust. Although Y2K is a neutral event, God may use it to spark a revival unlike any other revival in history. Our job is to repent, seek God's face and align ourselves with His ultimate purposes.

We really see the ultimate in hypocrisy coming from nearly every Christian author who is espousing the theory of Y2K as God's judgment upon us, with part of that judgment due to our reliance on technology and making computers and technology our God. Nearly everyone spreading this message is using a computer to print it out. Or (please insert your tongue-in-cheek as you read this next line) even more "wicked", they are putting it on their web site on of all evil tools-the Internet!

2. *"Those of us connected to the world's power grid will be the first to feel the effect of God's judgment."* First of all, there is a group of people who believe all man's technology is evil; the Amish. Unless you believe an agrarian lifestyle, building your own furniture, making all your own clothes, and growing all your own food is, in a sense, a more Godly way of life, then you need not feel "guilty" for being hooked up to the local power company. If we choose to remain on the power grid through January 1, 2000, or anytime for that matter, we must take responsibility for ensuring our families' well-being in the chance of a blackout. This is true regardless of the Y2K event! However, staying where God has placed you, as salt in a thirsty world, can offer unprecedented witnessing and evangelizing opportunities.

Skeptics claim Christians have completely failed at being salt in their communities and why would tough times motivate us to evangelize when we never did before. We would say to these critics, HOGWASH. Many sincere, faithful, Bible-believing, Christ-honoring families fight the good fight everyday in every city and in every neighborhood. In our church alone, families minister to the lost, visit the sick and elderly, provide extraordinary support to missionaries, send out our own missionary families to Mexico, Russia, the Philippines, and elsewhere, rebuild broken marriages and heal family hurts. Let the Lord use you wherever you are. If you need to be in the country, let God clearly demonstrate your need to move. But let it be to further His Kingdom, not to run away in fear.

3. *"All Christians should flee to the countryside."* Many who propose this use the Prophet Isaiah as Biblical support to this recommendation. While the "idea" of sitting on the porch of your handmade log cabin, sipping tea and watching the sunset, is so terribly romantic, the harsh reality is very un-romantic. We for one, have no desire to go back 100 years back; back to chopping all your own wood, sewing all your own clothes, tending broken bones with homemade splints, treating illness with homemade remedies, growing and processing all your own food. In essence, you will spend all your waking hours *just* ensuring your daily survival. Most people have neither

the skills, nor tools to accomplish this. Before you flee the city (and every opportunity to minister and witness) and re-locate to a rural area, carefully count the costs. Why are you doing it? Is it of fear? How will you provide an income? Who will you fellowship with? Is your wife or husband in 100% agreement? Are you leaving elderly grandparents behind? Are Christians called to be the salt and light of the world or should they all hightail it to the woods? We cover this area extensively in Chapter 14.

4. *"Send your money now to get our ultimate Y2K survival kit, book, MRE's, food storage, investment package, tape set, yada yada yada...."* Keep in mind that almost everyone who wants to teach you how to handle the Y2K crisis has an agenda. They want to make money off you. One author advertised his $89 set of tapes repeatedly in WORLD magazine. We ordered the tapes, listened to them. We then found a copy of his book, which contained essentially the identical information as the tapes, for only $20 on the Internet. We thought, why would he advertise an expensive set of tapes to Christians but sell a much cheaper hardcover book to anyone at Amazon.com? Only he can answer that question. We ordered a $189 set of material advertised as a Y2K survival kit. Very little in the kit was about Y2K. Most covered general survival tips, gold and silver buying recommendations, and bartering tips. While the information was helpful, we did not consider it worth $189.

It seems those with the loudest cry of "the end is near," have something to sell you. As a writer in the San Francisco Examiner pointed out, "Each consumer's $300 is a business's $3 million. Thus it should come as no surprise that many of the 'industry analysts' who foretell Y2K doom also sell Y2K 'solutions.'"[8] Be wary of those who tell you the only way to safety and security is to buy their exclusive product. The Gartner Group, a Connecticut-based IT (Information Technology) consultant firm whose profits rely primarily on marketing pricey Y2K advice to corporate clients, speculates that over $600 billion will change hands globally on Y2K compliance. Everyone wants to quote the Gartner Group when they want to scare someone. But keep in mind, their primary business is helping companies overcome their Y2K vulnerabilities. "Enter Gartner's recently expanded set of Y2K consulting services, including research accounts priced at $20,000 a year. Gartner's Web site (www.gartner.com) peddles books and brochures, such as a report on the growth in the IT services market-place. Titled 'Fixing Year 2000,' it's 12 pages long and costs $795."[9]

Other researchers disagree with the doom and gloom that Gartner is predicting." The U.S. Y2K problem is badly overstated," agreed Tom Oleson, research director at IDC Research, a market research firm in Framingham, Mass. "The companies selling Y2K solutions tend to exaggerate the problem to attract customers for their services or seminars."[10]

What is most disconcerting is when we see Christians picking up on these erroneous doomsday prophecies and spouting them as truth. We have seen it in many publications. For example, *The Coming Home Y2K Newslette*r quotes Daniel Fisher as saying "Like the regional power outage of the Northeast in the 1965, and the recent outage of the Northwest, (caused by faulty transmission lines in Oregon) power failures occur in a cascading manner. By causing unexpected bursts, shortages, miscalculations, and other bugs throughout the system, it only takes a minor communication error or weak link to shut down otherwise functioning plants. When large numbers of grid participants fail, they must draw power from the remaining grid to reboot their systems. This takes up to six times nor-mal power."[11]

Later, under the Utilities section, we will show you this is a total-ly inaccurate picture of how the power grid works. We must be care-ful about publishing these types of statements without checking their validity. Similar quotes or prophecies show up everywhere. It is all unsubstantiated and secondhand. The problem comes when people trust these sources and radically alter their lifestyle by moving out to a remote area and/or spend 1,000's of dollars on preparedness items due to inaccurate information.

5. *"Time is running out and the end times are near."* We have heard Christians claiming these current times are just like Joseph in Egypt. Store up during the years of plenty to provide for the lean times. Although we feel this is generally a sound Biblical *strategy*, it does not necessarily indicate the endtimes. Jesus said the end times would be like as unto "Noah" when all the citizens of the earth were eating and drinking (one big party) when the disaster hits. Now we realize that many in our world think they are in one big party right now. BUT given the prophets of doom Y2K forecasts- that the world

will be slowing descending into chaos- it will not be a sudden event. On the other hand, there are many people who are not participating in the current world's party. They live a meager subsistence, barely surviving in depraved, wicked, and vile circumstances. Take the example of Africa. Currently, millions are being destroyed through rampant AIDS due to immorality. Millions more are being slaughtered through ethnic genocide. Believe me, not much partying goin' on there!

During the time of Noah, all the people lived immoral lifestyles but faithful Noah was instructed to build his ark right where he lived! We believe God can lead people to move to a rural area but most people are not prepared for that type of lifestyle and/or unable to move for whatever reason.

Other Biblical examples exist contrary to the "flee to the hills" strategy:

- Nehemiah was told to go to the destroyed city, live there, and rebuild the city with a tool in one hand and a sword in the other.

- Daniel was kidnapped from his native land. His parents chose to remain in a city that was being conquered by enemies of their faith. Everyone knows how Daniel influenced even the wicked leaders of his time. God used him in a might way.

- Shadrach, Meschach, and Abednego. Ditto for these guys.

The bottom line message: Do not move out to the country in haste, fear, or hysteria. These are not "spiritual" reasons. Obey God, not your fear.

But Y2K is a Fact!

However, the Y2K crisis is REAL! Choosing to ignore it, choosing to hope a bunch of computer programmers can fix it, choosing to think it won't affect you is like choosing to ignore the a fireman when he says to get out of the building NOW! Your making the choice to ignore it or dismiss it won't stop the fire and it won't protect you, not matter how wrong you think that fireman is. Ladies and Gentleman, the fire has already started. It started years ago (as we just explained.) And one of the most startling things about this fire is that we will start feeling the heat far sooner then January 1, 2000. The real damage might really be starting to be felt as soon as the summer or fall of 1999. Although it is a minority of software applications affected, the potential impact could be very large.

And the clock keeps ticking....

Disclaimer:

Previously we wrote that many of the people telling us to buy gold and silver were not very credible sources of advice. Initially we realize that we had some misunderstandings and misconceptions about this whole Year 2000 computer problem. Let us explain what it was which turned us around on this issue. Basically, as we did our research, we started to get the facts instead of forming opinions based on limited information and limited understanding.

Something we really want to stress is this next point. The following few chapters have a lot of potentially depressing doom and gloom in them. These are all realistic possibilities of what can and will happen if each one is not dealt with and fixed on an individual basis. However, NOT all these things WILL happen. Many of these problems *will* get fixed. Many will get fixed 100% before they cause any of the described problems. Others will get mostly fixed. Some will get fixed shortly after they start causing problems. Some will only get partially fixed. And some may produce long term problems that take many months (possibly even years) to get fixed. At the time of publishing this Handbook, NO ONE can say which or how much. The Y2K problem is like a moving target. Most doomsday analysts are only looking at a current snapshot. By looking at snapshot of the problem now, of course we would agree that the problem is likely to be a catastrophe. However, the Y2K landscape is *continually changing,* always improving. Yes, we are running out of time but many things will be fixed in time.

However, this is our "full disclaimer" of the following few chapters: It is our opinion (and we will be the first to admit we can be wrong, either overestimating or underestimating) that most of these potential problems will not happen or will only have minor effects. Yes, some will happen. And it is possible that most of them could *potentially* happen. There is no way to know which or how bad they will be come January or February 2000. What we firmly believe is that problems related to the Year 2000 computer glitch will occur and they will cause in some way, at the very least, some disruption of every life in the civilized world. But we DO NOT think these disruptions will cause ANY of the following situations, as many others are predicting:

- We will NOT have the collapse of the value of the dollar.
- We will NOT have the collapse of the banking system.
- We will NOT have the total collapse of Wall Street.
- We will NOT have the collapse of the Federal (or State or Local) government.
- We will NOT have no electricity.
  (double negative intended)
- We will NOT have a nuclear meltdown.
- We will NOT lose our fresh water supply.
- We will NOT have total anarchy and major civil unrest..
- We will NOT have the end of modern civilization as we know it.

We just do not believe these sorts of major catastrophes are going to happen. We will explain why we do not think these things are going to happen. But there will be some problems. We do believe Year 2000 computer related problems will bring on the following:

- World wide recession.
- Drop (but not a total collapse) of Stock Markets around the world.
- Unemployment and layoffs.
- Increased business & personal bankruptcies internationally.
- Drop in real estate values.
- Sporadic shortages of various consumer products.
- Possible civil unrest, definitely internationally.
- Potential economic and currency instability internationally.
- Potential political instability internationally.

Yes, these are our *opinions* of what we think will not happen and what will happen as a result of the Year 2000 computer problem. However the computer glitch itself has nothing to do with opinions, feeling, or reason. It is a very real, unquestionable, scientific fact, with absolute certainty of its existence. On the other hand, there are varying opinions as to how bad it will get because no one has the ability to predict *how much of the Year 2000 computer problem can be fixed between now and January 1, 2000.*

Experts or Kooks?
Who are the ones "Crying Wolf"?

Leading experts from every facet of the computer industry, government, and business have recognized the potential dangers of Y2K.

Capers Jones is a noted computer engineer and an author of a book quantifying the costs and consequences of the Y2K problem. He says, "When the twentieth century ends, many software applications will either stop working or produce erroneous results since their logic cannot accept the transition from 1999 to 2000, when dates change from 99 to 00."[12]

Well-known Christian family finances expert Larry Burkett says, "I'm now convinced that Y2K-related problems could well pose the most serious threat to our economy since the Great Depression of the 1930s."[13]

One of the most vocal government official is Representative Steve Horn. He has spearheaded congressional hearings and issues a quarterly report card on the progress of Y2K work within government agencies. In his latest report, dated November 23, 1998, he says, "In a little over 400 days, America and the world will come to terms with a problem of our own making. The problem is described in various ways – the Year 2000 problem, the millennium bug, the "Y2K" problem."[14] The U.S. Government's General Accounting Office says, "The public faces a high risk that critical services provided by the government and the private sector could be severely disrupted by the Year 2000 computer crisis. Financial transactions could be delayed, flights grounded, power lost, and national defense affected."[15]

And what does the government say about all this?

- Just because government officials are saying it will happen doesn't mean it won't happen.

- Just because government officials are saying it won't happen doesn't mean it won't happen.

If that sounds a little confusing, it is because it is confusing. Yes, there are government leaders who acknowledge the Y2K problem and its potential severity. The reason for the first part of the headline ("And just because government officials are saying it *will* happen doesn't mean it *won't* happen.") is because so many of us have grown so use to not believing the government anymore. Well, this next statement will shock a few readers, but the government isn't always wrong. On the other hand, the government is often wrong.

In this case, both statements are true. In general, our Government (from local to national) acknowledges the problem exists. They back up that acknowledgment to the tune of billions of dollars already spent working on Y2K-related problems with their computer systems. Total future budgeted expenditures are in the tens of billions. In fact, "Joel Willemssen of the General Accounting Office said the government is not where it needs to be and, as a result, there will be important computing system failures. The government's Y2K costs have increased to $7.2 billion (three times the initial estimate of $2.3) and that figure is expected to grow."[16] Yes, there are those who would contend that our government does spend billions and billions of dollars on problems that don't really exist. We would agree with some of that. But in this case, you need to see this spending as an indication of them knowing something is serious enough to spend billions of dollars to try and fix it. No matter how you look at it, no matter what your personal opinion of the Y2K issue, you can't call this insignificant.

As for the second part of the above contradictory statement ("And just because government officials are saying it won't happen doesn't mean it won't happen.") we say there is much confusion within the government about the Y2K problem. What we mean by the government saying "it won't happen" is explained in the following statement. Many think the government will be able to fix its Y2K problems before it becomes a problem, but we say this is overly optimistic thinking. Yes, some areas of the government will be able to weather

January 2000 with little or no problem. But there are too many agencies, which started working on the problem to late too be able to fix it before the clock strikes midnight on December 31, 1999.

All indicators show Social Security will be fine. This Federal Agency started on this problem about the earliest of any government agency. However there are many mixed reports as to the status of other governmental agencies. The one that gives most "Doomsdayers" the worst fears is the Welfare department. Some are predicting welfare checks will not be able to go out on schedule in January 2000, resulting in riots and civil unrest in major metropolitan areas with a high welfare recipient population. We cannot tell you if the checks will go out on schedule or not. Even the government probably cannot truthfully tell you at this time. We just know they are working on it. But if the regular flow of welfare checks gets interrupted too long, it would be hard to imagine civil unrest *not* happening.

Probably one of the most tragic things about our government's response to the Y2K issue it the almost total lack of response to it by elected officials, especially those at the federal level. Nearly every issue on Capitol Hill and at the White House takes a higher priority than Y2K takes for politicians. Relative to the efforts of the business sector, our Federally elected leaders are doing very little. It is our prediction that when the general public starts suffering from various Y2K related problems, there will be a massive outcry to hold these politicians accountable for their lack of action. Yes, they are holding committee meetings and passing a few Y2K–related bills. But compared to the magnitude of the potential problems here, their actions have been pretty minimal. We have far more confidence in the business world to handle its Y2K problems than in the government. Businesses such as banks, credit card companies, retailers, manufacturers, etc., know their very existence is at stake. Of those who do not assess their Y2K vulnerability and remedy the problem now will most likely fail. If Widget Manufacturing's processes do not work on Monday, January 3, 2000, another company will step in and take their market share. It is called a free market system. However, if AFDC has errant computers who do not send out checks, there is no other alternative. The government has a monopoly on these services it provides and therefore has no great motivation to protect its "market share." In addition, its customers have nowhere else to turn.

We will go into more details in later chapters about the specifics of potential problems the government might be facing.

Major Fallacies involving Y2K

Myth: "I don't see why it is such a big deal."
Our Response: Read Chapter 2.
Myth: "Why does it matter? Can't they just change the clocks?"
Our Response: To change the clocks would require nearly the same computing manpower as it does to rewrite the date code.
Myth: "I don't have a computer, so it won't affect me."
Our Response: Your personal computer will be the least of your worries. The potential effects within industries, governments, and utilities will ripple across the entire population.
Myth: "They will get it all fixed in time."
Our Response: No doubt many essential software applications will be fixed in time. However, some systems will not. The sheer magnitude of the Y2K problem will prevent 100% repair completion. Computer programs in use by large corporations, government agencies and banks are enormous in their sheer size. LOC or lines of code is the term most often used to quantify these applications. A consumer software application such as Microsoft Word for Windows may have about 100,000 lines of code. Many software packages in use by corporations can top 250,000 LOC, while major systems may run up to 10 million lines of code or greater. While not every line of code needs to be repaired or even checked, the programming involved in Y2K conversion is astronomical. Caper Jones estimates there are approximately 1,702,125,000 function points needing Y2K work. (Function points are the lines of code that involve actual programming functions.)

Another factor affecting the success of Y2K conversion is the programming axiom that fixing one problem can create an entire set of new problems. Programmers must fix code and then re-fix any corrupted code.

Even if there was time to repair all systems, there just aren't enough programmers. Many of these systems were created using COBOL and other high level programming languages. COBOL and other languages have not been in demand as much anymore so there are few programmers (only about 550,000 trained in COBOL) trained in these languages available now. Already, any programmer with COBOL or similar Y2K conversion skills is demanding and getting $100,000 per year salaries. Private firms are hiring away any skilled government programmers with lucrative bonuses up front. In summary, the year 2000 problem is likely to be the most labor intensive and therefore expensive set of software changes and updates ever undertaken in history. Estimates run at about $1-$3 per line of code. Roughly translated, this is approximately $300–$600 billion.[17]

Lastly, obviously, we are just running out of time. The year 2000 will come exactly on schedule and some date repairs will not be finished. Also, some of the date repairs will not even have been tested to ensure they will work post–Y2K. Most software experts plan for at least half of the repair time be devoted entirely to testing the rewritten code.

Myth: "They can replace old software with new software."

While many companies will do just that, many others will not have the time and resources to replace all non-compliant systems. Many non-compliant computer systems within large banks, government agencies etc., are so big that creating new systems takes years of software development, big budgets and years of manpower. According to Caper Jones, "None of the major software applications in either the United States or the rest of the world can be replaced between now and the rest of the century. You have to fix the year 2000 problem in your current applications, like it or not." In addition, there are many kinds of specialized software for which there is no realistic replacement.[18]

Myth: It's "the-end-of-the-world-as-we-know-it"
(or TEOTWAWKI)

This addresses the other side of the Y2K issue. The first side, as we just mentioned, is all the fallacies and misunderstandings that Y2K will not have any impact or influence on anyone's lives. On the other hand, there are those who are predicting that the ultimate impact of Y2K will result in "It's-the-end-of-the-world-as-we-know-it" That is just going too far at the other extreme. What is most likely to happen will be somewhere in the middle, as we will be covering in this Handbook.

CHAPTER 2

WHAT HAPPENS WHEN THE CLOCK TURNS 12:00 AM ON 1/1/2000

(Authors' note: What we will be covering in these next few sections must be examined and analyzed by the reader from a very specific viewpoint as you go through the following. What we will be covering here is all based on scientific, computer facts–with as little opinion as possible. So the reader needs to absorb this from the perspective of fact, not theory, if they are to be able to make wise, informed decisions in response to the coming of Y2K)

Some before, some after, all related to 1/1/2000

In this chapter, we will discuss some various *specific* examples of possible effects and then cover each part of our computerized world and discuss their Y2K exposure and their readiness.

Just what can happen at midnight?

In the computer world, reliable, accurate information is the key to successful computing. If your input information, your data, is corrupted, then any outputted information is useless. When a program encounters information that has dates that need to be in the year 2000 and beyond, unless that computer program has been made "Y2K compliant," the program will fail or produce unreliable results. Often the program relies on the operating system, which is like the nerve system of the computer, to tell it an accurate time. But if the operating system is not Year 2000 compliant, the program will receive erroneous data. What makes the potential for disaster so ominous, is the fact that every industry, company, utility, government, business, school, hospital in the country uses computers for many tasks. Some of these tasks are mission critical to the day-to-day functions of these operations. If the computers fail, checks do not go out, lights do not come on, heat systems don't function, appointments are lost, medical records are lost, mortgages are calculated with errors, manufacturing systems fail, and so on.

But not everything will happen at midnight!

Although this chapter assumes that most Y2K–induced problems would occur at midnight, December 31, 1999, keep in mind that many Y2K–induced problems will be spread out over many months in the coming years. MSNBC news reports that, "Indeed, only 8 % of all date-related errors will hit on Jan. 1, 2000, according to the Gartner Group, which believes the majority of Y2K errors will strike over the next three years in relatively equal portions. After 2001, the problems will sporadically continue to strike as "dormant code" in legacy applications occasionally triggers errors."[19]

This is actually a good thing!

Because every Y2K–induced glitch will not strike simultaneously, it gives computer engineers and others a chance to handle each problem one at a time rather than be overwhelmed with massive, multiple failures. It also gives a good perspective on how similar systems may react. If one type of banking program fails in July 1999 (The start of the fiscal year for some companies and states), then others may need checking to avert similar problems. In fact three notorious dates have already passed with barely a whimper: July 1, 1998, October 1, 1998, and January 1, 1999. The MSNBC report continues, "Two key dates have already passed without major incident: the start of fiscal years for 46 states on July 1, 1998, and for the federal government on October 1, 1998. On each date, government computers began to look forward into fiscal year 2000 to perform projections and calculate benefits. Errors were expected, and no significant interruption of government services occurred. But those were warm-ups for the main event, and observers will study several critical dates to gauge how computer systems respond to the errors."

Now that January 1, 1999 has passed, it is extremely notable that although Y2K problems did crop up, they were of a minor nature, were fixed quickly and did not affect any other systems.

Embedded Chips

One major problem most people are unaware of is *embedded chips.* Tiny computer chips called microchips have been placed in almost every electronic device created. These microchips are in your car, microwave, VCR, coffee maker, TV, digital radio, and many other electronic devices right in your own home. What does "embedded" mean? Ed Yourdon, author of *Time Bomb 2000*, explains it this way "Historically, the term was used within the computer industry to describe a small "micro-computer" that was literally embedded within some larger piece of engineering equipment or industrial product. The embedded system provided the intelligence..." to control processes within different types of electronically controlled systems. These chips can be found in any device that has built in computer logic."[20] Some chips have a clock function which helps it process information. Some of these, about 2%, have a year date function and may have a year 2000 rollover problem. What does this have to do with Y2K? Several things:

- These microchips are very difficult to find and sometimes even more difficult to replace.

- There are billions of these chips in place right now. 3.5 billion microprocessors were sold in 1995, and 7 billion were sold in 1997. By now, there are probably at least 30 billion of these little guys floating around.[21]

- A small percentage of these chips are likely to be year sensitive and a small percentage of these could be described as "mission critical", meaning if they fail, it would be a bad thing. David Hall, a year 2000 consultant, estimates that "embedded systems will ultimately account for as much as 80% of the final year 2000 bill."[22]

It is the sheer numbers of these embedded chips that makes these repairs or updates nearly insurmountable. However, keep in mind that most embedded chips are NOT date sensitive and that the year 2000 will not affect them. If you have electronic devices you suspect will not be year 2000 compliant, namely, devices that have a date/time function, be prepared to replace most of these devices. The real challenge is finding, testing and replacing the embedded chips that are not in the home but in the workplace. Most will not be tested by 1/1/2000 and a small percentage will fail. Repair of these chips will take weeks if not months. As we cover different industries that are Y2K vulnerable, we will mention the impact of embedded systems and chips have their Y2K readiness.

Risk analysis eliminates quite a few embedded chips. For example, microchips in cars do not have a battery backup, display or microprocessor.[23] Therefore, they have little risk of calendar–induced failure.

Repair of embedded chips that do have the potential of failure is slow and difficult. There are several factors slowing this repair time:

• A technician must be available to come to the site. Since more than one system will blank out simultaneously, scheduling a repair may sound like this:

"Hello, the elevator in our office building is not leaving the basement. Can you send out a repairman?"

"Yes, I can schedule you for, umm, sometime in May. May 13th. Is a.m. or p.m. better for you?'

Although this situation will be rare, it will happen and it will need to be fixed.

• The technician must be able to locate the errant embedded system. Some embedded systems are sealed in underground bank vaults, offshore drilling rigs, in satellites, etc.

- Lastly, a replacement chip must be available. Some systems are too old and the manufacturer may not support it or it is just too expensive to replace the chip compared with buying a new device.[24]

For some embedded chips, some experts are recommending just turning it off for the date rollover. David Eddy points out, "For devices which are controlled by embedded chips that are counting the passage of time, simply having the device turned off as it crosses the 1/1/00 line may indeed be the best and safest thing to do, assuming you can easily get to the device to turn it off and then back on again."[25]

In fact, in some power plants, this is already being done during daylight savings time changes to allow systems to rollover to a new time without causing trouble. Another option is to remove "the batteries of the real-time clocks to introduce a reset and allow some level of functioning."[26] Embedded chip experts recommend considering going "to a safe mode for the year change- a low production or idle mode, with a high state of alert. So if a problem does crop up, you can deal with it easily. A mode where you know what to do with it if something goes wrong." says Bill Tarallo, systems integrator at Process Control Consultants.[27] Lastly, embedded chip experts point out that, "mission-critical systems are designed specifically not to be dependent on any one embedded component or software routine."[28] This reduces the risks of Y2K induced failures caused by errant embedded chips.

But will my VCR work on January 1, 2000? According to one source, "the only home consumer electronics most people need to worry about are most newer VCRs and older computers; the VCRs because only the new ones keep track of what year it is."[29] Test your VCR yourself; set the date ahead to Dec. 31, 1999, 11:58 p.m. if it has a year function; don't worry about it if it does not. Most other electronic products such as coffee makers, washing machines, and microwaves may have clocks but do not have calendar functions.

Utilities

Utilities are the big one. This is the one industry that doomsdayers hinge their whole end of the world scenario. And they are right about one thing. IF, the entire power grid goes down and almost all of us lose our power for a long period of time, the end of the whole world would be a very accurate description. However, the utilities' Y2K vulnerability is not as high risk as most doomsdayers would like you to believe. We have done a lot of research on how electric companies are physically set up and how computers affect their functions. We feel very confident that most of us will have power on January 1, 2000. There is a possibility that there will be *localized, sporadic,* power outages or blackouts of short duration but the utility companies will get the lights on within a reasonable time.

In this next section, we will often quote a power plant expert named Dick Mills. Here is a short biography from his web site: "Dick Mills has been creating software for power plants and power systems for more than 30 years. He was a pioneer in operator training simulators, helping five different companies get established. He has more than 2000 hours of simulator control room experience in emergency and startup conditions. He created the first independent network simulator, PTI's PSS; still the industry standard 25 years later. Dick also designed turbine controls, integrated plant controls, the automatic generation controls used in energy management systems, nuclear power plant probabilistic risk assessment tools, and process monitoring, archiving and optimization systems. Dick clicked to the Y2K problem in 1997. He switched careers to work full time on Y2K in power grids."[30] We feel he can give us an authoritative perspective on power plants and their Y2K vulnerability.

Here are some reasons why we are optimistic about the power grid:

Dick Mills says, "No matter how big the blackout, we can restore service to just about everyone within 24 to 72 hours. The operators can operate some misbehaving things manually and make do without some others. Some authors have predicted one month or longer blackouts because of Y2K. Poppycock! The power system protection systems are aimed at causing things to trip at most incongruities as a precaution so that equipment is protected from damage and power can be restored rapidly. This makes blackout events more frequent, but shorter in duration. Keeping things in proper perspective, a national blackout would not be that big a deal."[31]

If we have a blackout with many power plants shutting down, it is fairly easy to get them up and running again. Some of the more pessimistic writers claim that a power plant needs six times more power to start the plant than to run it. Their assertion is that if there is no power, there is no way to start up the plant again. Therefore, even the most minimal power shortage would cause long-term blackout. However, utilities expert Dick Mills says a plant only needs about 2% not 600% to "black-start" the plant. Also, "all 200 nuclear plants and many fossil fuel plants have diesel generators whose purpose is to supply the internal power needs in case of emergency." He relates one story from The Great Northeast Blackout of 1965. "At one plant, the operators had to burn office furniture to start a fire in the furnace, to boil the water, to make the steam, to spin the turbine, to turn the generator, to make the power. Amazing. The point is that they did it."[32] He says, bottom line, "a black grid can be restarted."

"Almost all of the 30,000 transmission substations in the United States are controlled by a series of dumb relays; the few that use microprocessors generally don't have date related action.... These controls only act like pressure-relief valves. They only protect the equipment; they do not control the flow of power.... Are power utilities handling their year 2000 problems? Yes.... Are all of the fixes installed? No. But they are on schedule to be installed by mid 1999. Is there going to be a global blackout? Very unlikely."[33]

"In the case of electric companies, date-coding plays only a minor role in the production of electricity, but it plays a crucial role in the metering of electricity use. 'They may have a Y2K problem, but you can bet that they are going to solve it because they want to charge and make money, which without working meters they cannot.' says Tom Oleson, research director at IDC Research, a market research firm in Framingham, Mass."[34]

The grid control computers power utilities use are EMS (Emergency Management Systems) and SCADA (Supervisory Control and Data Acquisition.) .Both of these systems, according to

Dick Mills, are not mission critical to the delivery of power. "...even if the SCADA and EMS systems are all broken, we can still get the job done [of restoring power.] It will take more hours to get it done without SCADA. Not months, or weeks, but hours."[35] If they fail, engineers are able to work around them. These two systems are the *only* systems that are dependent on telecommunications. Therefore, if these systems are not mission critical, then the availability of phone service is not mission critical.[36]

Generally power plants have 20 to 30% more capacity to generate power than needed to satisfy the demand. If some plants fail, we do have extra capacity to help compensate. Also, when the clock turns over to midnight, we will be entering a weekend in winter, typically not a peak electrical usage time.

Other doomsdayers are predicting massive blackouts if every single bug is not fixed by January 1. This is not true. Most plants have multiple pumps or generators with many opportunities for operator intervention in the event of failure. Dick Mills says, "...the widespread belief that all the Y2K problems in power system computers must be repaired before we can restore electric service are way off base."[37]

Lastly, let's put to death another myth floating around out there. Power outages or shortages WILL NOT cause your radios, TV's etc. to blow up. When a power shortage occurs, other generators increase their output to take up the slack. This results in small fluctuations in the frequency of the power flowing. "There have been numerous myths circulating on the Internet that frequency changes will cause motors to burn out and TV sets to explode. They aren't true."[38]

As for nuclear power plants, you can stop worrying. No meltdowns. Computers do not control most of the safety systems in nuclear power plants. "Most of the controls that operate the plant's reactor are electromechanical, which aren't date sensitive and so pose no Y2K threat."[39] Also, a vice president at the Nuclear Energy Institute claims that "the U.S. Nuclear Regulatory Commission and industry moved quickly to assess the possible risk of major year-2000 issues, and both have concluded that safety systems will, if required, safely shut down a plant."[40]

Most nuclear plants should be operating post Y2K. Here in Minnesota, the Nuclear Regulatory Commission (NRC) chose our

Monticello plant for its first ever Y2K audit of a nuclear plant. The NRC plans to perform Y2K audits at 12 nuclear power plants by January 1999. The NRC has found good progress and expects the plant to running come January 1, 2000. According to the Monticello plant's principal electrical engineer, nuclear power plants have few Y2K vulnerabilities. He says the auditors have found little to be concerned about even after inspecting 12 other plants.[41] Also, a vice president at the Nuclear Energy Institute claims that "nearly half [as of 10/5/98] of U.S. plants have completed detailed testing, certification, or correction of components and system affected by year-2000 issues. Most recently, the industry developed a contingency guidance manual to help plant operators prepare for potential offsite challenges to their plants."[42]

Our local utility, NSP, is planning to have extra generating capacity on-line on January 1 to prevent power shortages and blackouts. However, they do not expect problems.[43] In January, 1999, they announced they expect "to be able to produce twice as much electricity at its roughly 50 power plants as will be needed the night of December 31, 1999."[44] We would assume many other utilities would be planning similar contingency plans.

Another myth involves the Global Positioning Satellites. "Persistent discussions on the Internet claim that when we reach the date of GPS rollover on August 22, 1999, power systems will fail. No. Neither GPS nor computers are critical to the process. The actual synchronism mechanisms are simpler, more elegant, and more interesting than any computer chip application. They have to be. They've been in use since before 1880."[45]

Larry Burkett, of Christian Financial Ministries, called a utility company near his home and asked if they had a contingency plan in place. They said: "Absolutely so. We're going to do what we did 30 years ago, where if you have power off...we will run cheater wires. We physically go out there with a long piece of wire, we clip it on this cable and then we'll run over and clip it on this cable and we'll send you power down the line, if that's necessary."[46]

Many people predicting complete power grid collapse often quote this statement from David Hall, an embedded systems consultant at Cara Corporation: "Every test I have seen done on an electrical power plant has caused it to shut down. Period. I know of no plant or facility investigated to this date that has passed without Y2K problems." There are two problems with this quote. First of all, it is an old quote and the fact that power plants are testing equipment is a good sign that they are on their way to ensuring their Y2K conversion.

Secondly, as of today, "at least four major electricals have done significant on-line generation testing and they now have over 70 plants operating at staggered dates after 1/1/2000 with no intention of going back until possibly as late as 2003. Most of the plants were rolled on a schedule that involved setting before 9/9/99 a couple of days and letting roll as well as 12/31/99 and 2/28/2000. Although some problems with various versions of the software were encountered and in the worst case one plant lost two of its four display screens in the OPS center, in no case was there a trip that caused loss of generation. If the group I attended was a representative sample, I think maybe half of the utilities may plan to stagger their rollover dates early to spread the risk of the grid into smaller more-manageable chunks."[47] This is very encouraging. We cannot stress enough the importance of this fact: *there are power plants running with their clocks set in the Year 2000!*

We have found little information about natural gas delivery. Perhaps because it just won't be an issue. One good bit of information we found: "Two major natural gas vendors stated that the equipment that could fail is programmed to fail in the 'open' position, and that significant redundancy exists from the various wellheads all the way to the "city gates." They said the main risk is compressor stations [sic] that will probably have additional manning planned."[48]

Summary

In summary, please do not lay awake at night worrying about your lights going out on January 1. Yes, some people may lose power for a short period of time (less than 24 hours) some for maybe a little longer, but most of us will not. The most likely scenario will be power shortages with localized brownouts. This may become more common in the summer months with normal peak electricity demands. Most likely, urban areas will get priority over rural areas. It would be ironic if the rural isolationist will be sitting in the dark while the urban city dweller is warm and cozy with their lights on.

Businesses

Retail

In the retail world, shelves are stocked every 24 hours, sometimes even more often. To accomplish full shelves every shopping day with just the right amount of product, almost every store, be it grocery, clothes, department or electronics store, uses bar codes and computers to track every item as it comes in the freight door in the back and out the front door. Every transaction is computerized and processed. They operate on the "just in time" principle. This means they keep only the inventory needed for the next few days (sometimes even less) on hand at all times. Reordering of new supplies is done by computer, sometimes with a direct link to the supplier. High power computer systems link such stores as Target with such super manufacturers as Proctor and Gamble. Inventory information is constantly updated and re-supply shipments are made continually, ensuring full shelves at all times.

Accounting and payroll is also computerized. Retailers are spending millions to re-tool their systems. Wal-Mart alone is spending only 60 million but their spokesman claims they have been re-tooling all along. Other stores are spending even more. Sears estimates that it will have spent $143 million when all its Year 2000 work is finished. They expect to be done by mid-1999. We feel that most stores will operate in some fashion after January 1, 2000 but retail spending overall will probably suffer due to the recession we will most likely suffer in 1999 and/or 2000. Lastly, there could be regional shortages of some items or some stores may have trouble keeping certain items in stock.

Manufacturing

Manufacturing will have a hard time adjusting to post–Y2K primarily due to the interdependence of all their suppliers and the presence of embedded chips in factory processes. Since so many factories also operate on the "just in time" (JIT) principle, any weakness in the supply chain can severely interrupt daily production. These JIT shipments are totally controlled by computer. Computerized software

31

programs keep constant track of the entire inventory on hand. When a new order from a customer is logged into the computer, an automated request and manufacturing order are sent out to the factory floor, where more is scheduled to be made to replenish the inventory sold on the new order. Requests for more raw materials and various parts are also sent out automatically by the computer.

Every factory has multiple suppliers who in turn have their own suppliers. If any link in the chain of suppliers is not Y2K compliant, that link may fail. GM experienced this first hand when one of its main suppliers went on strike, resulting in numerous plant closings of their own. We once experienced this same thing in our own business as well. We had backorders of a particular horse software program that every supplier we called was out of stock on. We called the manufacturer and found the true source of the problem. In each software box, there was a little toy horse. Apparently the company who made the little toy horse was behind in its shipments. This resulted in the software company not being able to package its product; which resulted in the software program not getting to the wholesalers; which resulted in it not getting to our mail order company; which resulted in it not getting to our customers in time for Christmas gift giving! What a mess it was!

However, we have read of some companies initiating contingency plans of additional stockpiles at the end of the year. In addition, one wonderful feature of our free market system is that profit driven individuals will pounce on opportunities to supply needed parts or product to a demanding market. This is called supply and demand economics!

Overall

Corporations are budgeting hundreds of billions to rewrite their software and replace hardware systems. Triaxsys Research published a study which "finds that the top 250 corporations alone will spend at least $37 billion to make sure their computers still work in the next century."[49]

Some feel that the overall effect of Y2K could have some benefits as well for business. Rick Cowles, an embedded systems expert, mentions three benefits:

1. "Total inventory of all systems, source code, and hard-

ware. How many companies really, truly, know where all of the skeletons are located? And even more, how many have it documented?? Y2K is obviously the perfect reason to develop that inventory and maintain it once complete.

2. "Retirement of systems and hardware that are no longer used or useful (or maybe there's some other system or hardware that's 10 times more efficient at doing the task). Again, Y2K is the perfect opportunity to retire systems or hardware that aren't being used anymore and just sucking up space and systems resources.

3. "Competitive advantage. The company that is truly, honestly, Y2K compliant is going to have a whopper of a business advantage over its competitor's who are not ready."[50]

Another consulting firm gives us its top ten Y2K benefits: (Or put another way...)

Y2K is a Good Thing!

1. Competitive Edge

2. Cheaper Insurance. With a strong Y2K compliance program, business risk insurance should be easier and cheaper to acquire. Banks and insurers are requiring or considering requiring a Y2K plan or proof of compliance.

3. Faster Time To Market. A Y2K plan will ensure availability of product.

4. Clearer Big Picture. The Y2K conversion process helps business leaders to evaluate their organization.

5. Better Documentation. Technology managers will have a better inventory of all software and hardware documentation (Documentation is usually used to describe the "manual(s)" that accompanies programs.) than ever before.

6. Getting Rid of Dead Wood. Management can eliminate

inefficient or useless systems.

7. Cleaner systems. By converting existing software systems line by line, organizations will clean out old software code and write new code and programs will run leaner and meaner.

8. Improved Testing Process. Future software will benefit from testing and quality assurance policies put in place for Y2K conversion.

9. Easier Maintenance. Companies can now use standard date routines, which are easier to maintain or modify.

10. More Detailed Mapping. The Y2K conversion will give programmers detailed information about every line of code they use.[51]

Even Larry Burkett, of Christian Financial Concepts (www.cfcministry.org), says, "I believe that Y2K will cause a mini boom in 1999, as companies discover that their just-in-time inventory system is not going to work. ...I believe that they'll start laying in additional inventories in the year 1999. Utilities companies will lay–in additional coal supplies because the rail system will not be particularly dependable. Other companies are going to lay in raw materials. Other companies are going to lay in finished materials. They're going to increase their inventory."[52]

Let me add a few other benefits:

Another economist is predicting no recession at all due to the increased spending of Y2K. Already computer companies are experiencing record setting growth, profits and stock prices primarily due to all the Y2K conversion work being done. Businesses are purchasing new computers, new systems, new software, consulting, and other new hardware. Even the Post Office is completely replacing its outdated system for a whole new one this year.

Overall, companies who are Y2K compliant will have a SIGNIFICANT advantage over those who are not and will most likely prosper in the coming years rather than suffer. Shaunti Feldhahn, in her book, *Y2K: The Millennium Bug*, points out economic research that shows, "35 percent [of businesses] would thrive, because they would fix their systems well in advance and use that as a serious competitive advantage."[53]

Banks

Since computers heavily control the banking industry, it is easy to see why Y2K can affect their entire existence. ATMs, mortgages, loans, check processing are all handled by very sophisticated and large software applications. Some of these systems were first installed in the 1950's when most programming included two digit year fields. In addition, banks have massive software applications to convert to Y2K, some banks with 200- 400 million program instructions. Lastly, the banks are all interconnected, with *2 trillion* dollars in electronic payments, securities and checks flowing back and forth 24 hours a day, 7 days a week, 365 days a year. Ed Yourdon explains, "But a potentially larger problem exists when banks communicate with one another– e.g., when your employer gives you a paycheck drawn on ABC bank and you deposit it in you account at XYZ bank, the two banks have to communicate to accomplish the shift of funds from ABC to XYZ. Even if all of the software in ABC is Year 200 compliant, problems could occur if XYZ's software is not Year-2000-compliant."[54] If these interbank transactions are hampered by Y2K glitches, check clearing times could lengthen considerably.

But keep in mind that banks are under strict regulations requiring very strict compliance to Federal Reserve rules. Even now the Fed is bearing down on all banks and other financial institutions, requiring them to report all Y2K plans and strategies. It's like being pregnant; you're either Y2K compliant or you're not. And if you are not, the Fed will not allow you to be in business. Period. "We put our systems through software addressing every line of code to satisfy audits and users," says Tony Jasinski, a systems department manager at Wells Fargo Bank.[55] Your best bet is to have your money in a larger bank and/or spread out among several different banks. Most larger banks have their Y2K plans well underway and will be ready come January 1. A new report from the Federal Deposit Insurance Corp., "shows that of the 10,092 banks it regulates, 94% are making satisfactory progress."[56] The Comptroller of the Currency estimates that about

80% of its audited banks were making satisfactory progress. Says Lou Marcoccio, research director of Gartner Group consultants, "...people seem to think banking is more vulnerable [to Y2K] than other industries. I don't think it's true."[57]

Larry Burkett talked to a banker friend of his who owns a bank, "What are you going to do? He said, 'well, I think we're going to be ready. Everything we've checked out says we're going to be ready, but you just never know. So for three days, in three different times in 1999, we're going to operate our entire bank manually. We're going to do nothing but fill out manual forms like we used to. We're going to file deposits manually. We're going to use calculators. We're not going to use computers. And we're going to see if it works...' That's called a contingency plan."[58]

Bank Panic?

Although most banks seem to making sufficient progress and have plenty of regulators breathing down their neck, an even more serious problem could face the banking industry even as soon as summer of 1999. If the public *perceives* that the banks are Y2K vulnerable and that their deposits are at risk, people may run to the bank and withdraw their money. Now, we am not going to go into a long discussion on the failings of our fractional reserve system and what it really is all about. But suffice it to say, there is far less money (cash) in the bank than people have on deposit. Why? Because, when you deposit your paycheck for example, the bank immediately loans it out to get interest on the balance before you start spending it. You may think, "Gee, I spend my paycheck practically before I get it. The bank only has my money for only hours!" This may be true, but multiply that by millions of depositors and millions of dollars and millions of hours and you can see how the banks make billions of dollars every year by *not* keeping your money sitting in their bank. Therefore, when crowds of depositors demand their money, there is insufficient *cash* on hand to satisfy these depositors. A panic sets in and then *everyone* is demanding all their money at all the banks. This is what happened in the 30's and is called a "run on the bank." When this happens, the banking regulatory agencies would most likely declare a bank holiday and close banks for several days while the panic subsides. Another action would entail limiting cash withdrawals to only $50/

day or some such amount. The banks simply need to let depositors know their Y2K conversion plans and progress. A smart bank would advertise its Y2K compliance and promote Y2K awareness to calm the crowds.

The Fed's Plans

In anticipation of this new demand for cash, the Federal Reserve is already, *for the first time ever*, planning to boost currency reserves by $50 billion to $150 billion, in addition to the $460 billion already in circulation. The Fed is estimating that each of the 70 million households will, on the average, withdraw $450 to pay for necessities, such as food and gas. Other contingency plans include adding workers at the Fed banks, printing larger denomination bills such as $50 or $20 instead of $5 or $10 bills, and allow older bills to remain in circulation.[59] This should alleviate any problems with localized bank runs of a short duration, but would be insufficient if everyone wanted all their money out. This is why we recommend procuring a cash reserve well before the end of 1999. See Chapter 9 for preparedness tips and cash strategies. Lastly, Federal Reserve Governor Edward Kelly mentions backup plans the Fed has in place: "We are not totally dependent upon electronic systems. The Fed still has paper systems that can serve as backups if there should, indeed, be some sort of failure on a large-scale basis."[60]

Credit Cards

Although some non-Y2K compliant credit cards got big publicity, our research and experience shows credit card companies are by far the most likely financial institution to be Y2K compliant. They have their very operation and lifeblood at stake. In our own business, we take credit cards with 00 or higher in the year date all the time and have never had a problem. We always get paid.

Banking Summary

The banks are making good progress and bigger banks offer better security because the Federal Reserve deems them "to big to fail." In other words, the consequences of one of these larger banks failing outweigh the cost of bailing them out. Therefore, the Fed will do whatever it can to keep these banks solvent and operating. "FIRST, my feel is that the financial services industry (banking, stock markets, and insurance) at least here in the US and probably in much of the English speaking world (Canada, UK, Australia, and New Zealand) is further along in their individual and collective Y2K efforts than many other industries. Here a variety of obscure federal regulators have been issuing a steady stream of directives to their charges. The tone of these communications is astonishingly and reassuringly blunt. The message is basically 'We're going to be watching you like a hawk. If we get even a whiff that you're not taking this Y2K thing seriously, we'll cut you up in little pieces and feed you to a healthy organization.'"[61] The SEC (Securities Exchange Commission) is "requiring companies to include Y2K information in their publicly available SEC filings, and the federal Good Samaritan law encourages firms to share Y2K information."[62]

Finance and Investment Firms

Investment companies use computers extensively for all essential operations. Every single stock trade is done electronically with millions of shares changing hands daily. Multitudes of computer systems from a myriad of investment and stockbroker firms are continuously communicating. Any Y2K glitch among them could cause a cascading disruption of millions of transactions. Accounts would get charged wrong, interest payments get fouled, and more. However, just like banks, financial and investment firms such as brokerages are under strict regulations. Already the Securities and Exchange Commission has charged 37 brokerage firms with failing to report in time on their year-2000 readiness as required by its rules. About half had to pay fines of up to $25,000.[63] Newsweek reports that "It's not uncommon to find gung-ho efforts like the one at Merrill Lynch: an 80-person Y2K division working in shifts, 24 hours a day, seven days a week. It'll cost the company $200 million, a sum that could hire Michael Eisner and fire Mike Ovitz. "Our return on investment is zero," says senior VP Howard Sorgen. "This will just enable us to stay in business."[64] To continue their existence, these firms will be either forced to be compliant or forced out of business.

In addition, Wall Street itself ran several Y2K tests this summer with good success. Another major securities firm, The Options Clearing Corporation, has already successfully tested all its software in a post Y2K mode. This particular firm clears all buy-and-sell contracts for 146 of the largest brokers in the US and abroad. To the leaders of this firm, "failure is not an option."[65] These firms know their future depends on Y2K compliance. They are working to upgrade all computers in all their processes. Another report shows this firm is running three sets of computer systems running in parallel, has back-up generators in place, can get power from two separate power grids, and has bought all essential personnel digital cell phones that can double as walkie talkies.[66] Now that's serious contingency planning!

An aside: On Wall Street, the computers that track the Dow Jones Industrial Average (DJIA) will have their own problems independent of Y2K: the DJ10K. When financial services developed their information systems, the programmers allotted only a four digit numeric field for the DJIA. As you know, the Dow Jones is hovering around in the mid–9000 range right now. If it hits 10000, these computers will have trouble with erroneous computer results. "Probably not one in 100 programs have a five digit DJIA field," says Richard Dalin, a COBOL programmer.[67]

Health Care

As medical techniques evolve, they become even more dependent on computerized functions. Potential Y2K problems include patient monitoring devices malfunction, operating room support systems disruption, medical instruments malfunction, patient billing records in error, and medical insurance billing in error.[68] Hospitals and doctors are diverting more time and resources to Y2K conversion projects and hope to be ready. "...Hospitals have announced aggressive Y2K plans."[69] Home medical devices such as heart monitors, pacemakers, and infusion pumps don't have calendar functions. "Gary Thompson, director of information services at Scripps Health, which runs several hospitals and clinics, has found that most medical devices less than seven years old will present no problems."[70]

Insurance

Insurance companies rely heavily upon immense computer systems to track policy due dates, benefits and interest calculations, annuities, payment records and more. Many of these systems have been running since the 50's and 60's and are very vulnerable to Y2K problems. Any Y2K glitches would severely affect the functioning of insurance companies. Therefore, the insurance industry as a whole has been actively completing Y2K conversion and is making good progress. However, the industry as a whole will suffer massive losses due to Y2K induced lawsuits. Picture it this way, your company gets sued. You call on your litigation insurance (if you have any) to handle the expense. We will cover the enormous specter of Y2K induced lawsuits in a different section.

Transportation

Within the transportation sector, many facets are Y2K vulnerable. Air traffic controls, airline schedules, embedded chips in cars, train schedules, fuel deliveries, and more are all controlled by computers to some degree. Any malfunction in any of these systems would cause them to behave erratically or not at all.

Cars

While many Y2K doomsdayers are predicting most vehicles will be inoperable in 2000, we feel that most will run fine. Yes, there are computer chips in cars. But for the most part they do not have internal clocks. The car companies promise the cars will keep running. "Car computers don't care what date it is."[71] If they have a continuously running clock (called a Real Time Clock or RTC in computer terms), this clock must have its own power source. Otherwise, when you remove the main battery in your car, this clock would stop working and reset itself when you replace your car battery. If this clock has its own power source, you would replace it every 3-4 years as part of the car's maintenance. With my computer, we must replace its battery after 3-4 years or the computer goes wacko. [Yes, that is a computer-technical term: "wacko." Just kidding :)] Every source we have read can only speculate that cars will have trouble; no one knows for sure.

Planes

In the airline industry, the FAA is implementing strict Y2K compliance on all air traffic controls and airplane systems. Systems that do not pass will be indefinitely grounded and/or replaced. Although we do not recommend flying on New Year's Eve or Day, we do not foresee planes falling out of the sky at midnight. "We're still in the assessment stage, determining how big the problem is," says Denies DeGaetano of the Federal Aviation Administration. One possible danger is computer lockup: while planes will keep moving at 12:01 a.m. on Jan. 1, 2000, the screens monitoring them, if not upgraded, might lock. Or the computers might know where the planes were, but mix them up with flights recorded at the same time on a previous day. "You can bet we're going to fix it," says DeGaetano.[72]

Boeing's Seattle-based head of Y2K programs, Mary Jean Olsen, has found herself repeating the phrase: "Computers do not fly planes, people do." She says the public should "absolutely not" be concerned about stepping onto its planes. "We have studied thousands of Boeing's and Boeing suppliers' systems and only three are sensitive to date rollover, and they will not compromise the safety of the flight."[73]

In fact, the head of the FAA plans on flying over the midnight hour through all the time zones on December 31, 1999 just to prove her faith in the system's Y2K compliance. "Federal Aviation Administration Director Jane Garvey said "she plans to board an airplane shortly before midnight December 31, 1999 and fly across the country. The FAA is spending $191 million to ensure that air traffic systems are ready."[74]

Trains

Railroad schedules are heavily dependent on computers to coordinate the millions of shipments trains carry every year. The railroad industry is working on their Y2K conversion and is making progress. Norfolk Southern Corp., a $4 billion railroad hauling and trucking company, is several months ahead of schedule and is planning "to complete its Y2K conversion by its third quarter of this year

[1998]."[75] This is encouraging! However, railroad-switching systems rely heavily on embedded chips. Multiple failures within the railroad distribution system would severely disrupt the flow of goods all across our country. Shortages of food and fuels would be a serious problem.

Governmental Agencies and their Operations

Local

Most likely, some local government computer systems will fail along with some automated systems like traffic lights. However, at the local level, response time is minimal and things will be fixed quicker. The scope of the problem will be smaller and easier to manage. Lastly, local governments are able to revert to a paper and pencil status quicker and easier than our massive federal bureaucracy.

State

Similar problems and solutions as the local scenario but many states have been slow to assess their Y2K compliance and implement strategies. In addition, some of the more populated states have nearly the headache the federal government has in reprogramming their software. In the most likely scenario, some but not all state-funded programs will be shut down or operating at minimal capacity. Some of the larger states, California, Pennsylvania, and Florida for example, are making very good progress.

Federal

This is the scariest problem and even under the best of conditions, no one has high hopes for our federal government to be Y2K compliant by Jan. 1, 2000. Recently there has been a flurry of congressional rumblings about getting Y2K compliant but we feel many government services will be seriously affected. Welfare checks may not go out. Food stamps unavailable, etc.

Transfer payments: "Transfer payment" is just a catchall phrase for all the money the government takes from the taxpayers and hands out to the non-taxpayers. It is a term that defines payments to individuals for which no current goods or services are exchanged; e.g., welfare payments, social security, AFDC, government grants, unemployment benefits, etc. A computer somewhere generates every check. Most doomsdayers are predicting these government checks will stop and welfare recipients will burn cities to the ground, raping and pillaging every neighborhood. This is a rather extreme prediction. Note that word, *Prediction*. No one knows for sure what will happen. However, this is our prediction. Some checks will be delayed. Agencies will muddle through, writing out checks by hand if necessary. Churches will step in to assist the needy for short–term needs. Some are suggesting unique solutions such as printing checks beforehand, or setting up direct deposit contingency plans.

David Eddy, a software specialist, suggests "What's wrong with agencies like the Social Security Administration, backed by the full faith and credit of the Federal Government arranging ahead of time for banks to continue to issue and honor deposit instructions that have actually been frozen at some point in time? Worst case is some dead people continue to get checks. Best case is that Aunt Millie, and millions like her, don't miss a check. The primary objective is to keep money circulating in the economy. The agencies and issuing banks can "run a tab" and settle up later after the 1/1/00 dust settles."[76]

IRS: The IRS has the worst collection of computer systems on the planet earth. We believe they will not function barely if not at all in 2000. Some stats on the IRS:

- Their software portfolio consists of 50,000 computer programs with 100 million program instructions, about 1/4 the size of the larger banks.

- In 1997, the IRS abandoned a $4 billion computer overhaul that was headed for failure.

- They have started later than most government agencies and with fewer programmers.

- They estimate their Y2K costs to be well over $600 million.[77]

However, what most people don't realize is that the IRS does not collect taxes. Your employer does. With every paycheck your employer pulls out the proper amount for payroll taxes and then deposits it into the US Treasury. The IRS only "enforces" this voluntary compliance by monitoring all payroll taxes, deductions, etc. If the IRS collapses, money will probably still continue to flow into government coffers via the payroll tax (and other taxes). If you decide to stop paying taxes somehow, hoping the IRS will never know due to wrecked computers, be aware that someday they will research all payroll records and deposits and track down all non-payers. If things are really bad, it is possible that our government will devise a quick and dirty method to continue to collect money from its citizens. For example, congress could establish a national sales tax. However, this may be a good thing.

Defense: Although many Y2K doomsdayers predict the immobilization of our national defenses, we feel confident the DoD's Y2K preparedness will be sufficient. However, the challenges the military faces are significant. First, keep in mind that the sheer amount of software applications the military uses is far more "than any other federal or state organization- more, even, than entire industries in the private sector; current estimates of the aggregate military software portfolio are in the range of 30 billion program instructions.[78] Also, many of these software applications are written in old, arcane computer languages that have few available programmers. Some of this software is in embedded systems such as weapons. Many weapons will have to be replaced if they cannot be reprogrammed. This will take many man-hours and much money.

The DoD also has many computer systems just managing its massive human organization with payroll record, insurance, retirement information, etc. Computers are the very heart of our technology-based military. Just recently, Representative Steve Horn's quarterly reports gave the DoD only a D on its report card. However, extensive Y2K conversion plans are underway and the DoD predicts most mission critical systems will be ready. My brother was Chief Petty Officer on the *USS Independence* (until July 1998) in charge of making and maintaining weapons and weapons systems. He says there is no problem and they are ready.

Political

This is the biggest unknown. How will our government react to real or even perceived Y2K problems? We refer to this as the "x factor" because it has the potential to actually have the greatest impact. The other half of the "x factor" is how the general public will react. "A person is smart, people are stupid." As a group, people can behave very illogically. Oftentimes, the most stupid government policies are a response to groups of people acting stupid.

Food rationing and confiscation

If enough hysteria is created in the mass media, we may begin to see signs of hoarding in 1999. This could result in spot shortages of certain items. In most emergency management literature, martial law and confiscation of personal property is recommended for severe emergencies. Some emergency agencies have outlined contingency plans that include rationing and confiscation of food supplies even in the event of *perceived* hoarding. Chuck Lanza is the head of the emergency team in Miami. He has dealt with many disasters in the area including Hurricane Andrew. Here is an excerpt from Chuck Lanza's plans for Y2K preparedness:

    I. Identify threats
        A. Pre-event hoarding
        B. Post-event shortages
        C. Run on markets and food distribution centers
    II. Develop threat-monitoring procedures

        A. Pre-event hoarding
            • Establish liaison with major grocery chains
            for early warning of hoarding
        B. Post-event shortages
            • Establish liaison with major food distributors
            to obtain periodic local food supply inventory

        C. Run on markets and food distribution centers
            • Establish liaison with major supermarkets for
            early warning of runs on markets and distribu
            tors

III. Identify actions that may eliminate risks in advance

    A. Pre-event hoarding:
- Public Education
- Voluntary rationing plan
- Mandatory rationing plan

    B. Post-event shortage
- Establish liaison with food producers in local area in order to gain control of local food inventory if necessary

    C. Run on markets and food distribution centers
- Develop a "market protection plan" with local law enforcement agencies
- Inform public of "market protection plan"

IV. Identify actions that may be taken to minimize the impact of risks that may materialize.

    A. Pre-event hoarding:
- Public Education
- Voluntary rationing plan
- Mandatory rationing plan

    B. Post-event shortage: Establish liaison with food producers in local area in order to gain control of local food inventory if necessary

    C. Run on markets and food distribution centers
- Develop a "market protection plan" with local law enforcement agencies
- Inform public of "market protection plan"[79]

Note that the recommended "gaining control" of food supplies is primarily a response to real or even *perceived* "hoarding" by the general public. Those citizens accused of "hoarding" may have their supplies confiscated and "re-distributed" to those in greater need. Although, as a Christian, we have no objection to sharing those in need, we have a serious problem with forced redistribution. This would be a far more dangerous situation than a few power shortages or computer crashes.

Election Year 2000

Don't forget we elect a new president in the year 2000. As we finish this Handbook, congress has impeached President Bill Clinton for perjury and other offenses. We will not go into the debate over Bill Clinton's presidency, but we examine the possible outcomes and its ramifications on the year 2000 scenarios.

Scenario One: Bill Clinton resigns or is removed from office and Al Gore becomes president. Our own prediction is that he may appoint Hillary Clinton as vice president to honor Clinton's "wonderful legacy" of liberal do-gooderism. Hillary's popularity polls are at an all time high right now. There is a chance that Al Gore may also face criminal charges as well. The year 2000 comes, bad things happen, and our high tech guru Al Gore is blamed for the mess. He could institute martial law and expand the government's power over our lives. He does not get re-elected. Ed Yardeni, senior economist with Deutsche Morgan Granule predicts that, "the Y2K explosion will blow up Mr. Gore's political bridge. The odds are higher that he will be swimming in the River Kwai than sitting in the Oval Office in 2001."[80]

Scenario Two: Bill Clinton resigns or is removed from office and Al Gore becomes president. He may appoint Hillary Clinton as vice president. The year 2000 comes, nothing bad happens, and our high-tech guru Al Gore is given credit for "fixing the mess." He could get re-elected.

Scenario Three: Bill Clinton stays in office. The year 2000 comes, bad things happen, and maybe Bill is blamed for the mess. However, due to the massive government computer problems, he suspends the elections, declares martial law, expands the government's power over our lives and clings to the presidency past the year 2000.

Scenario Four: Bill Clinton stays in office. The year 2000 comes, nothing really bad happens, and Bill finishes his term uneventfully. The Democrats find someone other than Al Gore to run in 2000.

Regardless of who is in office, the potential exists for martial law to be enacted. Grant R. Jeffrey, a noted end–times author points out, "The range of almost dictatorial powers available to a U.S. president after he declares a national emergency equals the vast legal powers held by Adolph Hitler during his dictatorship in Germany."[81] There are some national security executive orders available to the president

47

if he declares a national emergency. Here are a few:

- The seizure of all print and electronic media- this could include the Internet.
- The seizure of food supplies and resources, public and private, including farms.
- Basically the seizure of everything in the U.S.

FEMA would have broad powers to control emergency efforts, currency flow, American financial institutions, and more. Of course, this is a very unlikely scenario but the possibility does exist. The government tends to take the attitude and approach that we, the general public, are too stupid to know what's best for ourselves, thus making it necessary for the government to "intercede" on our behalf, for our own good and well being.

One last scenario: one world government. Now, normally, we don't entertain thoughts and anxieties concerning one-world government conspiracies. However, the Y2K crisis is unique. Nothing in history has had the potential to be a "economic, political and military crisis of such vast proportions that no nation, on its own, could possibly solve it. The Y2K computer crisis provides a unique opportunity to impose a global government solution."[82] Many nations are even further behind than the US in their Y2K conversion process. Grant Jeffrey quotes David Rockefeller, one of the key leaders of the Council on Foreign Relations, as saying, "We are on the verge of a global transformation. All we need is the right major crisis and the nations will accept the New World Order."[83] Jeffrey speculates that the UN will appoint a Y2K Czar to inspect and enforce worldwide Y2K conversion. Stiff penalties would be levied against non-compliant countries. The Bible predicts a one-world government. It will happen someday. Whether it will happen now is entirely up to God. One thing for sure: now, more than ever before, is the time to repent and preach the gospel.

Litigation-lawsuits

If you thought Y2K conversions sound expensive, wait till the lawyers get involved! Consider this scenario. You have your life savings in the stock of XYZ corporation. This corporation is a good sound company with a record of performing well. However, its CEO fails to recognize the threat of Y2K and does not allocate sufficient time and resources to get its systems completely Y2K compliant. On January 4, 2000, the first business day of the millennium, the company runs into multitudes of Y2K failures, cannot operate adequately, and starts running losses. Its stock price plummets and you lose your life's savings. Not only do you lose; thousands of other stockholders lose. A greedy lawyer pounces on all of you and prepares a class action suit charging the company with negligence, lack of fiduciary responsibility (an obligation to act in the best interests of the shareholders of the company they serve), failure to act, etc. Capers Jones explains that CEOs and business leaders really have only 2 game options to consider:

> "If the year 2000 problem is not serious, then those of us who caution executives that it is serious will be embarrassed, but no real harm will occur other than the fact that authors of books like this one will look a bit foolish.

> If the year 2000 problem is very serious, and executives fail to take the warnings seriously and plan corrective actions, then they will be called into court on charges of violation of fiduciary duty. Indeed, if year 200 problems should cause accidental deaths or injuries, possible criminal charges might occur."[84]

Not only will stockholders sue their companies, business will sue each other for failure to provide their contracted services or materials and companies will sue software vendors for providing "defective" software. Even government agencies will get involved. Companies are sort of caught between a rock and a hard place. One newspaper article puts it this way: "Discussing Year 2000 activities is plenty uncomfortable for American companies, because they're trying to communicate separately with two audiences. One audience is the public, which wants to be told that things are okey-dokey. The other audience is filled with securities lawyers itching to sue companies that encounter disaster Jan. 1, 2000 and see their stocks crater Jan. 2." [85] Compound this scenario many times over, and you will understand why Capers Jones speculates the following. "It is possible that litigation expenses (and any damages if suits are lost) for the year 2000 problem can exceed the cost of [Y2K] repairs by as much as 20:1 when negligence and violation of fiduciary duty are proved or at least confirmed by jury decisions." [86] And of course, the only ones who will profit from all this suing and countersuing will be the lawyers themselves! The Giga Group estimates the year 2000 litigation costs as perhaps *topping one trillion dollars!*

One way to avoid this problem is for Congress to pass laws offering Y2K lawsuit protections, caps on Y2K damage suits, and relaxation of antitrust laws to facilitate the sharing of Y2K repair technologies. Already, President Clinton has signed a law designed to encourage U.S. Corporations to share Y2K information. Called the *Year 2000 Information and Readiness Disclosure Act*, it offers companies limited liability protection for pooling information about Y2K technologies, products, methods and best practices. "This legislation will help provide businesses, governments and other organizations with the necessary informational tools to overcome the Y2K computer problem," Clinton said. [87] Only time will tell if this will help.

In addition, some states such as California are considering legislation to limit or prevent consumers from asking courts for punitive damages, or payments for pain and suffering, related to year-2000 computer failures. "Some sort of reasonable public policy in this area has to be implemented, or we're going to put our high-tech industry at a severe disadvantage to computer companies in other nations," says an aide to a California legislator. We feel more legislation will come to pass to help out software companies. Even more strict legislation of a different kind is being considered in Britain. "A conservative member of Parliament introduced a new bill that would make it *illegal* for companies in Great Britain not to correct the year-2000 problem." [88] Although the intentions behind this bill are honorable, it is the equivalent to passing a law to make poverty illegal. While it contains good intentions, it just ain't gonna work!

Summary of the effects of Y2K

We will cover the different probable scenarios in a later chapter but we wanted to let you in on some other viewpoints. We have read many different opinions on Y2K and its possible outcome. One thing that continually comes to my attention is that many authors make wild statements as if they were absolute fact. Consider these very opposing viewpoints:

Two computer software experts' views:

"The Year 2000 people who have studied the Year 2000 problem are in agreement to a large extent about the consequences:

- There will be relatively few large incidents that attract international attention. Most elevators with embedded microprocessors will work just fine, but the news reports will be filled with the case of one elevator that grinds to a screeching halt. This will matter because in that elevator is - the Queen of England? The Pope? The President of Russia? You?

- There will be a number of problems affecting limited numbers of people. At least one bank may cease functioning for a period of at least a day. This won't make the news unless it's a very big bank; it won't matter to you unless it's your bank.

- An annoying large number of systems, computers, and appliances will call out for attention. Remember that VCRs, ovens, car readouts, and personal computers usually can't make it past twice-yearly daylight savings time adjustments without help. Why should they make it past December 31, 1999 without help."[89]

Contrast this with a different author (with whom we disagree):

- Social security checks will stop coming.
- Planes all over the world will be grounded.
- Credit card charges will be rejected
- Military defense systems will fail

- There will be massive, long-term power failures.
- Banks funds will be inaccessible"[90]

We have tried to provide accurate information and facts in our descriptions of the Y2K effects, not hyped–up speculations. Also, we have included opinions from real people who are experts in their field.

SECTION II

SCENARIOS

Introduction to Scenarios

No human alive has the ability to predict and tell you what is going to happen as a result of Y2K. Sure, it is easy to say that we could lose power, water, and most consumer goods if *everything is left "as-is"* and Y2K related problems are not dealt with. This is sort of like saying, "Gee, if you keep driving straight where the road curves up ahead, you will drive off the cliff." You do not have to have a Ph.D. to figure that out. As we researched this Handbook, the frustration we are having is that this is sort of the attitude many others seem to have. Almost like everyone running around saying, "The sky, is falling! The sky is falling! It won't get fixed! It can't get fixed! Head for the hills!"

But this attitude is totally wrong.

As already mentioned, no one can accurately predict what will actually happen because there are literally billions of variables to all this. No one can predict exactly which critical computer systems and embedded chips will be fixed or replaced between now and then.

Previously, we have written about individual effects by themselves, in an attempt to help you understand the full nature of the problem, We are now going to approach the Y2K issue form a different perspective.

What our goal is with this chapter is to give you a general overview of the combined effects Y2K could have on our lives.

We will be writing from the perspective of three different scenarios, which will give an overview of the combined effects Y2K might have. We will first give a best case scenario. Then we will describe what our best "guestimate" is of what we speculate is most likely to happen. Last, we will give you the "worst case scenario" which will probably be most similar to the picture being painted by most other authors writing about the effects of Y2K.

CHAPTER 3

BEST CASE SCENARIO

Best case scenario: an inconvenience

General overall picture for the best case scenario

There are several main areas that will feel the effects of Y2K: the "x factor," the economy, and the overall everyday life. Because there is no way to accurately predict with certainty what things will fail in what areas, good, basic, broad–based preparedness is still recommended even for the best of what we can expect to happen. There will be a complete section devoted to this later in the Handbook.

Y2K and the human element: the "x factor."

If January 1, 2000 comes, and for all practical purposes, very little actual Y2K related failures occur, Y2K will still have a definite, real effect on everyone, due to the human element.

We will start approaching this issue a little differently. Obviously, as of this writing, January 1, 2000 has not come yet. And as of this writing, there have been very few Y2K related failures that have actually occurred. Yes, this is for most of the obvious reasons; namely, Y2K has not "hit" yet.

And, of course, many things are already being done to try to prevent as many bad things from happening as possible. There are many things being done from a very methodical, planned, and predictable perspective. And then there is the unpredictable: **the human element—the "x factor."** In the next section, we will be covering more about the unknown human element as a variable to possible scenarios. We cannot stress this next point enough.

The human response element or this "x factor" will be a *key* factor as to how well our society makes it into the beginning of the next millennium. We feel the way people panic or not panic may have the biggest effect, even greater than any actual Y2K–induced problems.

We have seen the enemy and the enemy is us.

Some of you will remember a comic strip call "Pogo." It was more of an editorial type cartoon. The most famous line to come out of that whole cartoon series was, "We have seen the enemy and the enemy is us." Basically, a spin on the saying "We are our own worst enemy." As far as Y2K goes, we believe that when all is said and done and everyone looks back at Y2K, "we ourselves" will have been the source of the worst problems related to Y2K.

But in the best case scenario, people do not panic. Hysteria remains under control, and people make rational, levelheaded decisions. They make their Y2K preparedness plans in a methodical, organized manner. Most people place great faith in "the system" and its ability to solve this problem with the least amount of disruption and inconvenience. Also, communities band together and solve their problems and help their neighbors. History shows that when a major disaster hits, people, as a whole, tend to seek to help each other rather than harm each other. Sharing of basic goods and services, a sense of camaraderie, and a desire to serve community emerges.

This is all, of course, in the best case scenario. Problem is, this is not always the way people tend to behave. We will address more of that issue in the next main section. So, as such, we expect that in the best case scenario, the worst Y2K related problems we will face will not be due to Y2K and its direct related effects, but due to the unknown human element of what people and/or the government do in response to perceived Y2K problems.

This is where the Church can truly shine. We can offer help to those in need, share resources and information, organize community preparedness, and spend hours in repentance and prayer seeking God's will for our families. As Shaunti Feldhahn points out, "The more you know about how Y2K may apply to you, the stronger a tool you will be in God's hand to accomplish His purposes."[91]

It's the economy we need to watch

"The Market" can be self-fulfilling.

This sub-section is partly related to the previous sub-section, the unknown human element. As such, we will be covering it in greater detail in the next main section about what we believe to be the highest probability scenario.

One thing Wall Street professionals have always been aware of, to a small degree, is that "The Market" can sometimes tend to take on sort of a life of its own and respond almost in a self-fulfilling way. The market goes up, because the market is going up. On the other hand, the market will go down, because the market is going down. News that was good for the market at one time is bad for the market another time. As far as Y2K goes, if everything seems to be going along fine and no major Y2K disasters seem to be looming on the horizon, then the financial markets, namely Wall Street, will be responding primarily based on the traditional fundamentals that it should be responding to.

Presently, the Dow Jones Average is hovering at record levels in the mid–9000 range. It is our opinion that overpriced Internet stocks are one of the primary factors fueling this stock market boom. For those of you who do not follow the stock market, we will explain. Stocks such as Amazon.com (a mail–order Internet bookseller), American Online, (an Internet service provider) and Yahoo (a popular Internet search engine) are all trading at huge, unprecedented *multiples* of their true value relative to the value of the company and its sales. Amazon.com stock has exploded 900% this past year and they have yet to make any profits! All this is fueled by speculation and greed.

At some point, these stockholders will start taking profits by selling off their holdings. Once the prices begin to slide, panic selling could very easily set in from other stockholders. Fearing they would lose their own profits as well, these other stockholders could start dumping other portions of their stock portfolios. Once a market slide begins, inertia and panic take over. Every stock plummets: the good, the bad and the ugly. (Note: this is all our opinion and is subject to fallible, human error.) Subsequently, it is also our opinion that the market will possibly suffer a major correction this year and this downturn will be initiated by massive sell–offs in these currently "hot" Internet stocks. This sort of predicting is very subjective. Or maybe a better word to use is *speculative*. As such, there is really no way to predict this. You just have to be aware that the possibility exists so you can watch for it and respond accordingly.

Granted, our economy is about the healthiest it has ever been. However, much of our economic growth is fueled by the highest rate of consumer debt ever. Any slight downturn could precipitate a sharp increase in personal and corporate bankruptcies when people are just not able to maintain their current level of debt. **We urge extreme caution in investing any money in the stock market right now.**

So as far as Y2K, the best case scenario, and the stock market, well, little or nothing could possibly happen. But realistically, we feel a stock market correction and a recession are the most likely possibilities during this time. As such, you will probably want to position your investment strategy accordingly.

Recession

Our bottom line opinion: not *if* a recession will hit, but **how bad**. Even in the best case scenario. With the hundreds of billions of dollars spent on Y2K related fixes, and with the billions of hours of manpower and productivity lost, we cannot avoid some recessionary impact. *Unless the American economy is still able to go through an incredible economic boon otherwise*, the Y2K costs and lost productivity could result in a serious negative economic impact.

There is another factor that will magnify any negative recessionary impact Y2K has on the American (and world) economy. In even the best possible scenarios, at least some business will fail specifically due to Y2K related problems. Either they will be the ones with the direct Y2K failures, or one or more key suppliers will not be able to deliver essential parts or products required for the operation of the business. There is really very little argument over if businesses will have Y2K problems and failures. The real question is how many and how bad. It seems that not many economists are paying attention to the Year 2000 Problem. An exception is Dr. Ed Yardeni, Chief Economist of Deutsche Morgan Grenfell in New York. Besides testifying before Congress on the seriousness of the problem, he maintains pages on his web site on Y2K. He comments:

"The Year 2000 Problem (Y2K) is a very serious threat to the US economy. Indeed, it is bound to disrupt the entire global economy. If the disruptions are significant and widespread, then a global recession is possible. Such a worldwide recession could last at least 12 months starting in January 2000, and it could be as severe as the 1973-74 global recession. That downturn was caused by the OPEC oil crisis, which is a useful analogy for thinking about the potential economic consequences of Y2K. Just as oil is a vital resource for our global economy, so is information. If the supply of information is disrupted, many economic activities will be impaired, if not entirely halted."[92]

And then there is the compound effect. The recessionary effect of all the money and productivity lost will compound the economic effect that all the weaker businesses are already suffering. One economist predicts this best case scenario: "Speed Bump: This is the scenario that most news organizations seem to be expecting. Problems occur beginning in 1999, but because they're widely expected, things slow down and we go over the bump without real damage. Snafus happen and businesses apologize for the errors. The event becomes the national equivalent of tax filing day, with many headaches, much griping, but in the end, the problems are corrected and life returns to normal speed."[93] Business magazines have also predicted an economic slowdown as an almost certainty. Business Week reported: "According to a new analysis prepared for BUSINESS WEEK by Standard & Poor's DRI, the growth rate in 1999 will be 0.3 percentage points lower as companies divert resources to fix the problem. Then Y2K could cut half a percentage point off growth in 2000 and early 2001."[94]

Small problems in isolated sectors of the economy have the potential to cause a ripple effect through the entire world economy. One just has to look at the severe recession of 1973-1974 to see how a small crisis such as a foreign oil shortage could ripple through the economy. Even though foreign oil was only a tiny fraction of the Gross Domestic Product, energy costs affected nearly every industry. Richard Scurry, a former vice president at IBM and now a financial executive, says, "My concern with Y2K is that it will be relatively small business problem, but that–when it winds its way through the economy–it will multiply."[95]

We will cover more on this next topic in a little bit; but while we are on this point we wanted to emphasize it here. If you start hearing about many business failures due to Y2K problems, start watching very close. If too many companies start going under, the recessionary domino effect could be very devastating. This is actually our greatest Y2K fear. We believe that of all the real, likely–to–happen Y2K–related problems that will impact our lives, the economic effect will be the worst.

Certainly there are other potential problems that could result in worse effects, such as loss of power or no food distribution. But, in our opinion, those things are not as likely to happen. On the other hand, the economic impact is one of the most realistic, predictable problems that is likely to come out of all this. The ripple effect of all this could go many ways. The most likely effects are lost jobs, reduced wages, and various economic hardships. We are not going to go into all the various effects and problems associated with a recession. That is not within the focus or scope of this Handbook. Our goal is to simply point out these potential problems and make you aware of what we believe the potential impacts of Y2K will be.

Investments

Just a little bit of an investment tip here, which we will be covering in much greater detail in the *Your Money and Y2K* section. If we do have a severe recession, it is most likely to have a deflationary, rather than inflationary influence. This is NOT the ideal economic environment to be holding a lot of "hard money" asset investments like gold or silver. There could be commodities whose prices may go up. Shortages of oil imports would very likely escalate gas prices at the pump almost immediately.

On the other hand, when money is tight, "CASH is King." Just as in the Depression, those with cash in hand got incredible bargains. Stocks, real estate, hard goods; everything was cheap. On the other hand, we will agree that IF the government opens the floodgates of spending, eventually it could result in inflation. But that would be something more long term and further down the road.

What will not happen in the best case scenario.

The general consensus of most experts on Y2K is that even in the best case scenario, there will still be a number of Y2K related failures. These failures, at best, will be unpredictable, random, and regionally localized. The result of these failures will range from minor inconveniences for many to major headaches or crises for a small few. At best there will be no major systems failures. Those systems that do fail will tend to be localized, with most being fixed in a relatively short time frame (hours to days). As far as the specific Y2K–related failures, as stated many times, it is impossible to say which things will specifically fail and which will not. It is best to look at all the possible systems that could have problems and figure most will be OK and a small mixture will have various, regionally sporadic failures. If this sounds like an incredible vague and nebulous answer, it is because it IS an intentionally vague and nebulous answer! There are likely to be some spot shortages of various things. Even if there are relatively few Y2K–related failures in factories, just the specter of Y2K is likely to result in some level of hoarding, which could result in some spot shortages of some basic essential goods and commodities.

Just to repeat what we wrote earlier, *in the best case scenarios*, we **DO NOT** think Y2K disruptions will cause ANY of the following situations, as many others are predicting:

• No major worldwide depression.

• No shut–down of some utilities.

• No shutdown of any banks and ATMs.

• No major drop (but not a total collapse) of Stock Markets around the world.

- Not high unemployment and layoffs.
- No increase in business & personal bankruptcies internationally.
- No steep drop in real estate values.
- No widespread shortages of various consumer products.
- No widespread civil unrest, possibly domestically, definitely internationally.
- No significant economic and currency instability internationally.
- No significant political instability internationally.

This is our opinion of a *"best case scenario"* resulting from Y2K. It is a **general overview** and is not intended to explore significant details. Please keep all this in mind as we go into the next section of what we predict are likely to be the most probable effects of Y2K.

CHAPTER 4

HIGHEST PROBABILITY SCENARIO:

SOMEWHERE IN-BETWEEN

Our opinion

We would like to emphasize that these scenarios are opinions, and we acknowledge room for error, and room for circumstances beyond the human ability to foresee, which can potentially change our predicted outcome.

With that said, we would like to restate what we have said previously: Based on the entirety of all the research we have done regarding Y2K, we believe the economic effect of Y2K will be the event that will have greatest impact on everyone's lives. Even Larry Burkett from Christian Financial Concepts agrees, "The physical side of this is not as bad as the alarmists have represented, I believe. The economic side that we've been talking about before, in my opinion, is much worse."[96] Read Chapter 3 again to fully understand the likelihood of a recession. We are not economists and have no formal economic background ourselves. But we are basing our conclusions on all available information we could find, and the opinions of experts in the position to speak authoritatively.

We will first cover the recessionary influences of Y2K. Then we will cover more of the most likely "human element" effects of Y2K. Lastly, we will cover the most likely to happen problems directly caused by Y2K related failures.

Most likely Y2K related problems

We wish to apologize in advance for the appearance of seeming to be a little redundant here, but it really cannot be avoided. We have three different scenarios to cover here. In order to do proper justice to the topic, it requires us to address many of the same issues three different times, to give the explanation for EACH scenario. With that said, we will summarize with a repeat of what we stated earlier:

We just do not believe major "It's-the-end-of-the-world-as-we-know-it" catastrophes are going to happen. And we explained why we do not think these things are going to happen. But there will be some problems. Under the best case scenarios, we stated that these events would probably not occur. However, we are of the opinion that the best case scenario would be unlikely. We feel the actual events will be somewhere between the worst and best case scenarios. Hence, we do believe Year 2000 computer related problems *will* bring on the following:

- Worldwide recession of some degree.
- A drop (but not a total collapse) of Stock Markets around the world.
- Higher unemployment and layoffs.
- Increased business & personal bankruptcies internationally.
- A drop in real estate values.
- Sporadic shortages of various consumer products.

- •Sporadic, short-term, regionalized power outages or short-
  ages
- • Sporadic civil unrest, possibly domestically, definitely
  internationally.
- • Some economic and currency instability internationally.
- • Some political instability internationally.

Recession a certainty

There is a little joke in economics that goes something like this. If you took all the economists in the world and laid them out in a single line, they still could not agree. Another thing about economists, it is very hard for them to do very accurate predicting very far into the future. One of the reasons for this is because of the human factor. These events include major natural disasters, major problems in foreign economies (look at how the Asian and Russian economies affected the stock market in 1998, virtually no one was able to predict either one), changes in tax laws, political decisions, new laws, and on and on. There are usually just too many unknown variables to do accurate long term predicting of where the economy will go. Almost any economist would agree.

The key word to the limitations in the previous paragraph is: UNpredictable. There are some very key things about the Y2K situation that are predictable and as such, we can make some VERY accurate economic and investment forecasts based on them.

The following are things about Y2K that we know for *certain* will happen:

- • $300-500 billion will be spent trying to fix Y2K related
  problems.
- • $250 billion to 1 trillion dollars more will be spent on lit-
  igation, settlements, and, penalties.
- • Countless millions of man hours of productivity lost.
- • Massive amounts of lost productive and profitable
  research and development.
- • Some systems will fail.

Those are the "absolutely we know will happen" things. Then there is the highly probable speculation as to how people are most likely to react. Mentally, take yourself back to what we wrote in the previous chapter about Y2K causing a recession. It is not a question of if, but of how bad. Also, as we wrote earlier, the recessionary, economic effects are likely to be the thing that will have the greatest impact and influence in most people's lives. As such, making plans for a recession is *one of the main strategies* you need to have in your Y2K preparation planning. We will not go into very many details as to specifically what to do here, because we go into significant coverage of that in other sections.

Y2K is very likely to have a ripple effect on many different markets, all resulting in similar effects: business weakness, cutbacks in business, reduced consumer spending, layoffs, money pulled out of financial markets, and such. Each one will make the effect on the others worse. A multiplying factor, or as just mentioned, a domino effect. We simply cannot see any other picture. One economist predicts this scenario for a moderate Y2K effect: "**Slow Drag**: In this scenario, problems appear over time before and after the year 2000. As daily, weekly, monthly, and other periodically run programs encounter the problem, there is a constant drag on economic activities. Billions of dollars are spent correcting the problems. Everything done in 1999 that carries over to 2000 or done for the first time after 2000 will be problematic. Delays, errors, and decreased productivity diffuse through the economy. Lawsuits proliferate, further dragging the economy down. The effect is as significant as a major increase in tax rates or energy prices. The year 2000 problem results in a recession."[97] Dr. Ed Yardeni, Chief Economist of Deutsche Morgan Grenfell in New York, a large investment group, predicts "the probability of a resulting global recession at 30%, but has now increased that to 40%."[98] Even now, Dr. Yardeni's estimates of the probability of a recession are rising every month. Roger Ferguson, governor of the Federal Reserve warned that Y2K, like the Asian financial crisis, "can have unexpected spillover effects" on the worldwide economy" if it isn't corrected.[99] (Note he says *"if"* it is not corrected.)

Events that could alter the probability and/or severity of a recession:

This does not mean that we cannot be wrong. There may be factors that we have missed in all our research. There may be economic factors that do not exist as of the time of this writing. We will give you a couple of examples, so that you can be looking for some signs that could cancel out our recessionary fears. The Federal Reserve Board could drop interest rates and that would have a massive stimulus effect on the economy. On the other hand, the highly speculative world of futures trading is betting that the Fed will actually raise interest rates. "Now, some futures traders are betting that potential Y2K problems at banks might cause interest rates to rise next year." [100]

The Federal government could pass massive tax cuts resulting in a major economic boom (just look back at Reaganomics!) Or our Federal Government could pass a flat tax or National Sales Tax resulting in massive accounting savings to taxpayer and businesses nationwide. This would also result in millions of able bodied lawyers and accountants now free to actually work PRODUCTIVELY instead of doing bureaucratic number–crunching, form–filing, and paper–shuffling. We can say this from the perspective of having started, owned, and operated three businesses. We know what we are taking about in terms of lost productivity in the current system. (We are not saying this to advocate a flat tax or National Sales Tax. We are just using this as a "for instance" example of something that could possibly have a very good economic impact but we are simply not able to predict or account for at the time of this writing.)

And, of course, our economy could be robust enough to weather any Y2K disruptions and financial setbacks. Just as in Joseph's time, the Lord has given us years of blessings to help us through the years of hardship.

It goes without saying that it is unlikely any of these particular things could happen. But it is possible some unknown factor could keep us out of recession. "But as for me and my house," we are preparing for recession as part of our Y2K preparedness strategy. The financial markets will most likely get very hard hit in many areas, as mentioned in the previous chapter. Examine all your stocks, mutual funds, money market accounts, and real estate holdings very closely. Look at each individual investment in terms of how would a recession impact the value of each one. Lastly, the *most important step* in preparing for recession is to *get out of debt* as much as possible and get ready for possible work slow-downs or possibly even unemployment. More details in a later chapter.

Y2K and the human element.

As we said in chapter three, we see the human element as the other major effect Y2K will have on everyone's lives. The human element has the possibility of having a larger role in product shortages and unavailability that does actual Y2K related failures. Even the best manufacturing and distribution conditions are no match for sudden panic buying and hoarding. You need to look no further than the news stories in areas just before a hurricane hits to see how fast normally stocked stores are wiped clean of various goods. We will cover this issue in greater detail in the chapter on preparation.

And the human element or the "x factor" can affect the stock market as well. Just look at the history of the stock market and you can see how emotions can send the market on a wild roller coaster ride. If you compound recessionary economic news with Y2K related panicking, the potential for major sell-offs in the financial markets is significant. And financial panic tends to breed more financial panic and selling. Even before there are any actual Y2K–related problems that actually hit companies and business, the financial markets could be sent into a nosedive.

On the other hand, for some, this roller coaster will be a real investment opportunity. Whenever the stock market goes on one of these roller coaster rides, some very good stocks get sucked down with the bad. Those types of companies will turn out to be great investments for anyone who buys those stocks when the market is down. We are not saying this in terms of any investment recommendation. We are just pointing it out. If you are interested in investing in the stock market, or anything else for that matter, please read the chapters on *"Your Money and Y2K."* Bottom line: don't invest in anything that you don't do plenty of research on and are fully aware of all the risks and nuances of whatever investment vehicle you choose.

Civil Unrest

Another area of the human element is civil unrest. We believe there is a possibility for sporadic, isolated civil unrest in various places both domestically and internationally. However, in our opinion, the probability for major civil unrest domestically is very unlikely. Local police departments are already gearing up for having total readiness for any problems that might occur. We know of some departments that have canceled all vacations and requests for time off the last couple of weeks in December 1999 and the first few weeks of January 2000. "Los Angeles Police Department officials, taking no chances, have asked the mayor and City Council to set aside $4.5 million to keep as many as 300 more officers on patrol around the clock for a week before and after the turn of the century."[101] Remember that these people are professionals. They are not going to go into this unprepared or naive about potential problems of civil unrest or various problems which have the possibility of occurring.

What we are going to say next is complete speculation, but we ask you to think through our reasoning here. If there are any signs of civil unrest within the United States, we believe the government (whether is local, state, or federal does not matter) will declare strict curfews or even martial law before they allow any civil unrest to erupt or spread. The general public will *demand* martial law if there is virtually any civil unrest, even if it requires the full deployment of the National Guard!

So, as you can see here, we are at great odds with many of the authors who give they advice that people need to move out to the middle of nowhere to "protect themselves" from all the supposed civil unrest and chaos they predict will break out due to Y2K. We will cover more on this issue in another section. We will also cover more about Y2K civil unrest in the next chapter.

Real Estate

There are two significant factors and one minor factor that we believe will contribute to the decline of the real estate market due to Y2K related factors. These two factors are the weakness of the economy and those people selling homes to "escape the city" and move out to the country. Between the two factors, urban and suburban real estate prices can only go down.

Keep in mind that there are always those unknown events that cannot be predicted or projected for, such as we mentioned a few paragraphs ago. But barring something unknown and totally unpredictable happening, there is simply no way to go but down for the real estate market as a result of Y2K.

Lack of buyers

In any market, one of the most significant factors in the direction of the real estate market is basic supply and demand. When the number of buyers goes up, the general direction of the market tends to be up. And the obvious counter-effect is when there are more sellers then buyers, prices are weaker and it is a buyer's market.

Up until now, the real estate market has been going through one of the best markets ever. The economy has been great, interest rates have been low, and because many families have two incomes, Americans have generally been able to have greater extra disposable income then ever before. This has been combined with even more wealth from record levels of the stock market. As a result, many families have been using some of this newfound wealth to trade up to bigger, better, and more expensive homes. But this flurry of home buying activity has only been possible due to all the extra income and low interest rates. Consumer debt has been increasing as well. If the economy weakens, and "consumer confidence" weakens, people are much less likely to buy a bigger, more expensive home. Unless they *have* to move right then, the tendency will be towards waiting to move until the economy improves.

Panic sellers

As far as Y2K is concerned, we see a greater desire to sell and move away from urban areas. Already, rural Realtors are seeing an increase of buying from people desiring to flee urban areas due to Y2K panic and other reasons as well. With the advance of technology and its subsequent affordability, more workers than ever are able to earn a living without even leaving their home. All things being equal, we cannot envision more people wanting to be moving into the city relative to the amount of those moving out of the city for these reasons. Whereas we are not going to go so far as to claim there will be a significant flight from the city, it is very likely it could be enough to have a negative (rather than positive) impact on real estate values.

There is another Y2K factor that will also probably result in weakness in the real estate market. This component is actually likely to have an even more significant impact than the flight from the city. Although the vast majority of people are not likely to go as far as selling their homes to move out in the country, Y2K related fears are far more likely to delay or prevent many potential moves to bigger, more expensive homes. In general, in times of uncertainty, people tend to want to stay put.

Loan availability

There is another, less well known, factor that has a significant effect on real estate values. It is the actual ease of credit. Even if interest rates are low, there is no guarantee of actually getting approval for a loan. There are times when almost all a person needs to get a house loan is a job. At other times, especially when the money supply is tight or during poor economic periods, only those with the best credit can get anything more than a small loan. If personal bankruptcies and business loan defaults continue to rise, banks and credit card companies will be even more picky about who they loan money to. Currently, bankruptcies are at an all time high despite our booming economy and low joblessness. Y2K is very likely to put us into such a tight credit situation.

Summary

We believe that the three factors we have just described will combine to result in an actual drop in real estate values throughout most the nation. In some areas, where the real estate market is especially overpriced, there is the potential for some major drops of values and prices.

If you are in the market to buy or sell a home in the next couple of years, you will want to keep these factors in mind.

Computers

We think that most computer problems are either not going to be a significant issue or will be relatively easy to fix. Contrary to what many doomsday Y2K authors are reporting, our research indicates that most companies generally seem to be addressing and dealing with this issue. Literally every week there are new patches and fixes becoming available to take care of Y2K computer problems. Surveys show that most companies are quickly assessing their Y2K problems and initiating Y2K attack plans. Computer Reseller News reports that "More than 70 percent of the largest IT [information technology] users expect completion of Y2K testing by mid-1999. There is a tracking poll that "is one of the longest-running surveys to systematically monitor Year 2000 readiness. Covering IT directors and managers of 114 major corporations in 12 industrial sectors and 13 federal, state and local government agencies, it is conducted by Cap Gemini America LLC, a market leader in Year 2000 services." After surveying these companies, Cap Gemini estimates that "while firms expect significant new challenges in 1999, *98 percent have launched full-fledged Year 2000 strategies* (italics added) which have helped them assume greater control over spending and staffing concerns."[102] Even if these companies have Y2K glitches come year 2000, companies are resourceful and will muddle through somehow.

INDUSTRIES and SYSTEMS: Their most likely status

Utilities

The utilities are the most talked about when the Y2K doomsdayers begin forecasting utter collapse. In an earlier chapter, we discussed how Y2K could affect utilities. In the most likely scenario, electric companies will experience localized outages of short duration. Phone service will probably be somewhat worse but both should be fixed within a few weeks. Telecommunication carriers "are well under way to achieving compliance" "The carriers are on top of this better than any other segment," says Bill Allred, a year 2000 specialist.[103] Cap Gemini, a Y2K conversion company rates telecommuncations as having an 82.5% score in fixing Y2K code.

Businesses and Manufacturing

Manufacturing may, even under the best scenarios, suffer some major problems. The domino effect may paralyze some industries for short periods of time. However, in our free market society, opportunists will immediately pounce on any holes in the supply chain. Other sources will increase production and new sources will appear. This has happened before in many industries during other interruptions in the supply chain. However, there will be spot shortages of short duration of some items.

Retail

Wal–Mart has been at the Y2K conversion for years and is confident they will open their doors on January 2, 2000. However, some of their suppliers may have Y2K problems of their own. But Wal-Mart and other stores are in the business of making money. Somehow they will figure out how to get stuff on the shelves and how take your money. They may not have all 35 brands of shampoo but they will at least some on the shelf! Overall, the probability is high that there will be spot shortages of all types of goods. But it is unlikely that grocery stores will be completely empty.

Summary

The editor of Internet Week magazine says that his "conversations with organizations that are actively working on year 2000 issues incline me to believe that the total breakdown of civilization is not really the issue. In fact, most things to which we are accustomed to will still be available."[104] We agree. Most companies will get most things fixed and we will suffer only minor problems for a short duration. The companies, who do not get things fixed, will not survive. Some business managers are planning on dropping some of their vendors or suppliers who are not compliant. "Among the IT [information technology] managers responding to the poll...60% will be changing IT vendors and about half expect that some business partners may be history because of noncompliance issues"[105] They will just simply find new suppliers who are compliant.

Banks

Banks, on the other hand, will experience little difficulty with a few smaller banks being shut down by either the Federal Reserve Bank or the FDIC for not being Y2K compliant. Bigger banks are the furthest along in Y2K conversion and will most likely absorb the slack. You should be able to transact "business as usual" and banks will be open. Larry Burkett feels that, "Clearly the banks are either going to be compliant or they're going to be shut down or they're going to be sold. I mean there's no alternative there.... I would have some cash on hand, obviously, but I wouldn't worry about my bank."[106] However, there could be limits on how much cash one can withdrawal daily and stricter limits on loan availability.

Finance and Investments

In the most likely scenario, finance and investment companies will be operating fine. However the stock market will slump and some stockbrokers will be out of business due to the failure of high-risk investments and speculative stock options.

Health Care

If you can avoid being in a hospital on New Year's Eve, we would highly recommend it. However, we feel medical care will not suffer significantly.

Transportation

While many Y2K doomsdayers are predicting most vehicles will be inoperable in 2000, we feel that most will run fine. Yes, there are computer chips in cars. But for the most part they do not have internal clocks. In the airline industry, the FAA is implementing strict Y2K compliance on all air traffic controls and airplane systems. Systems that do not pass will be indefinitely grounded and/or replaced. Although we do not recommend flying on New Years Eve or Day, we do not foresee planes falling out of the sky at midnight.

The biggest effect that we will most likely experience is intermittent failure of the delivery systems of needed supplies. Shipment of ABC product won't get to City XYZ in time to fill the shelves at store PQR. The cause of this will be glitches in trucking and railroad companies computers. This is why you will want to stock up on food and supplies.

Many third world countries that supply our oil will not make their Y2K conversions by January 1, 2000. Dr. Harrison Fox, Professional Staff, U.S House of Representatives, subcommittee on Government Management, Information and Technology (Rep. Steven Horn's subcommittee) reported that, "Three of the five oil refineries in Venezuela (which provides some seventeen percent of the oil supply in the United States) will not be compliant in time and will have to be shut down. These and other problems will lead to disruptions in the flow of oil."[107] Subsequently, gasoline may be in short supply or at very high prices due to limited overseas oil imports. Tank up your cars before New Year's Eve and consider staying home for a few days.

Governmental Agencies and Operations

Local:

Most likely, some local government computer systems will fail along with some automated systems like traffic lights. However, at the local level, response time is normally minimal and things can be fixed quicker. The scope of the problem will be smaller and easier to manage. Lastly, local governments are able to revert to a paper and pencil status quicker and easier than our massive federal bureaucracy is.

State:

Similar problems and solutions as the local scenario but many states have been slow to assess their Y2K compliance and implement strategies. In addition, some of the more populated states have nearly the headache the federal government has in reprogramming their software. In the most likely scenario, some but not all state-funded programs will be shut down or operating at minimal capacity. Many people think, "this is a good thing."

Federal:

Most federal agencies will have most of their critical system conversion work done by January 1, 2000. However, there will still be some problems but the bureaucrats will muddle through. We suggest not depending on getting any federal checks other than a Social Security check.

IRS: the IRS has the worst collection of computer systems on the planet earth. We believe they will not function even under the best conditions, if not at all, in 2000.

Defense: Although many Y2K doomsdayers predict the immobilization of our national defenses, we feel confident the DoD will be sufficiently ready to continue defending our country.

Political

This is the biggest unknown. How will our government react to real or even perceived Y2K problems? In most emergency management literature, martial law and confiscation of personal property is recommended for severe emergencies. The highest probability is that the government will enact martial law in those areas hardest hit. Hard hit areas will include regions or towns where electricity may be off or food shortages and panic buying may be starting. However, "hard hit areas" will be scarce and most cities and towns will be operating normally. Election 2000 should be very interesting.

Litigation

This is going to the biggest headache for sure! There is no disagreement that the litigation costs from multitudes of Y2K disaster induced lawsuits will be astronomical. We do think the probability is high that the government will pass legislation to limit the amounts awarded in Y2K lawsuits. One lawsuit has already been settled in Michigan. "TEC America Inc. will pay $250,000 to Produce Palace International Inc., a Warren, Michigan, grocer, to settle the first lawsuit by a company claiming its computer systems wouldn't work because of year 2000 date-field problems."[108] The amount awarded however, is not much more than the amount they paid for the software; in other words, it was just a refund.

What will not happen

The potential for great chaos and disaster does exist, but we do not believe it will come to the worse case scenario, as described in the next section. One writer brought up an interesting perspective that we want you to think about. In the article, "Are we all going to die?", Howard Belasco says a resounding "No! We should treat Y2K as a hurricane or blizzard that will snarl our lives for a week or so, but the good news here is that because this is NOT a natural disaster, there will NOT be the necessity to rebuild or to repair. While industry will be effected, the physical infrastructure of the world remains intact. We will *still have the cities, farms, bridges, roads, and harbors, power lines, trees, buildings and tools.*"[109] (Italics added.) Some doomsdayers talk as if the landscape will be wiped clean and every bit of civilization obliterated. But we don't think so. Civilization will go on. At least until the Lord returns.

CHAPTER 5

WORST CASE SCENARIO

There are some people who believe the Y2K problem will be worse than the Great Depression. Whereas the potential for disaster is greater, we do not believe it will have an aftermath of that magnitude. We feel people should not *act* for the worse. But if they know the signs to be looking for, they will be able to see it coming sooner if it does come. And everyone needs to have a backup, a "Plan B", if the worse should occur.

We will discuss that "Plan B" in Chapter 6, after we briefly explain here what some see as the worst case scenario.

"The end of civilization as we know it."

We will not spend a lot of time on this section. There is not a lot of reason to. The needs that you would have, if the following things happen, go far beyond what this Handbook is designed for. Our primary goal here is not to prepare you for the worst, but give you an idea of what could potentially happen. If your desire is to prepare for the worst, we have the following thoughts:

Preparing for the worst would require you to make drastic, major changes to your lifestyle. It truly would involve selling everything (at least the house and a lot of your conveniences) and "heading for the hills" (or the woods, or the open prairie, or anywhere else as far from major cities as you can get.) It would require becoming self–sufficient in many areas such as power (electricity), fresh water, food, necessities, and anything else you may require for a year or two.

If you do take this approach, it goes without saying that if the worst does happen, you will be in the best position to get through it. But let's look at the other side of the coin here. If the worst does not happen, where will that put you? What will you have given up in making all your drastic preparations? The obvious things you would give up are job (income), home, and a lot of money. What we mean by a lot of money is that you will have had to spend significant amounts of money to make preparations unnecessarily. It will be extremely difficult to get back to where you were in regards of material things: car, house, furniture, and anything else you sell off or leave behind. Of course, in the eternal scheme of things, those things are only temporal. They are not needed by God to serve His purposes.

There will be something else that you may lose and will be extremely hard to regain, if it is even possible to regain, and that is your *witness*. Another area to consider is losing your *opportunity* to witness to all those people you currently know, as we do go though whatever we all go through. If all your planning turns out for nothing, then you most certainly will be labeled "one of those kooks" and probably a "religious kook" by all whom know you.

With all this said, we understand, and realize that some will want to do all this anyway. So for whatever reason, there are some people who feel they should leave "civilization" behind, and move out to a very rural setting. For those of you who feel that way, we just want to challenge you to make sure you have looked to God for guidance in this decision. Be sure you have a firm understanding on how moving to an extremely rural area is God's will for your life and how it can ultimately contribute to spreading the Gospel. The same could be said for moving anywhere. For those of you who are seriously considering trying to become self-sufficient, please read chapter 14, "Your *Church and Y2K.*"

Worst Case Scenarios Effects

(Any combination of the following:)

In the worst case scenario, most communities would be without electricity for long periods of time, 30 days or more. Water supplies would be in jeopardy, and food would be scarce. Cities would experience frequent episodes of civil unrest and riots. Cash would be the most important asset along with your stored foods and supplies. The stock market would plummet to a fraction of where is now and stay there for over a year. Retirement funds and lifetime investments would be devastated. Unemployment would reach record highs. There will be more frequent incidents of terrorism due to failed security systems. The Great Depression of the 30's would not seem so bad. We don't feel that the entire nation would go down in flames but things will be bad in the worst case. It would take months if not years to get most systems back on-line.

We could possibly survive all this if it were not for how the government may react to all this. Conspiracy lovers will finally see their predictions come true. Some things the government may do in reaction to these problems:

- Pass laws against hoarding and initiate search and seizure of private food supplies

- Recall all currency due to hyperinflation

- Institute martial law and curfews

- Suspend the 2000 elections

- Sign UN treaties to create one world government

We fear our own government could potentially bring on the worst problems due to Y2K. Whether or not it comes to that will depend more on what the Lord allows to happen than anything else.

CHAPTER 6

KEEP YOUR FINGER IN THE WIND! RIGHT NOW!

So what is the deal with this "Keep your finger in the wind! Right now!" title? Well, because that is exactly what you need to be doing. Figuratively speaking, you need to put your finger in the pre-Y2K wind and get a sense as to where this thing is headed. And start doing it right now! We will elaborate briefly in summary, and then we will get into the nitty-gritty detail.

What we mean by all of this is:

What you do NOT want to do is to simply do nothing at all. (Double negative intended)

What you also do NOT want to do is decide Y2K is coming, make a "to-do" checklist, complete your checklist, and sit back and wait.

This is where the "finger in the wind" part comes in, and we cannot emphasize the importance of all of this enough. Yes, you want to start making your list. But you need to have 2 lists. And that's not all. Your first list should be all the things you and your spouse (we don't do that "significant mate" or "partner" politically correct stuff here. If you haven't figured it out by now, the authors are "Politically Incorrect" and proud of it. Having 7 children and homeschooling, we aren't even allowed to be politically correct! Enough ranting and raving...for now!) Anyway, back to what we were saying.... You and your spouse need to agree on all the things you really need to do and/or change in your lives, families, homes, finances, etc. due to Y2K, and put that all down on a checklist. We will get back to this list in a moment.

The second list

This list needs to be your "Plan B" list, as previously discussed. Exactly what are you going to do if (pardon our language, but this next slang term really does seem the most appropriate) all hell breaks loose? No electricity. No banks. No water. No food or consumer goods in the stores. And add to all this, civil unrest. As we have already said, we do not think this is going to happen. But the important thing to remember is that any or all of these things COULD possibly happen. And you do need to have a well thought-out, viable plan of action in place for what you will do if any of these worst case scenario situations do occur. And this is where the "finger in the wind" comes into play.

What to be watching for

As you work towards completing your first list, you need to constantly be watching and listening for signs and sounds of the tremblings and rumblings of any of the worst case elements. Right now, we can't tell you exactly what all those signs might be. We can only give you some examples of what they might be. If any of the following situations occur, be watching *very* carefully, and look out!

Watch out for these Top 10 Y2K dates for Y2K troubles:

1. January 1, 1999. At the beginning of this year, testing is beginning for systems all over the world. It is notable that this date has now passed with very little fanfare. Although Y2K problems did surface, they were very minor, rare, and fixed quickly. Also, USA Today, (January 13, 1999) reported that, "A Y2K glitch in one place didn't spread through interconnected systems and zap other computers, as feared." The headline read, "Easy change to '99 squashes some Y2K doom and gloom."

2. April 1, 1999. On this date, Canada, Japan, and the State of New York begin their fiscal year.

3. July 1, 1999. On this date, forty-four U.S. states begin their fiscal years.

4. August 22, 1999. On this date, the Global Positioning Satellite (GPS) technology will fail in earthbound receivers that are not upgraded or replaced. The GPS system has twenty-four satellites that transmit signals to earth. These signals are picked up by electronic receivers to determine a vehicle's exact location and velocity. This would most likely result in problems routing traffic.[110] Many companies have replaced these receivers. A report from MSNBC comments, "Most commercial users of GPS have upgraded their systems. For example, the Federal Aviation Administration, a major user of GPS data, has upgraded its systems to handle the system rollover, as have airlines."[111]

5. September 9, 1999. Many doomsdayers predict this date may cause problems because the number "9999" was used to indicate "end of file," causing a program to come to a complete stop. However, this is not necessarily true. One computer specialist put it this way: "In our opinion, the Nines Problem is a massive red herring. Neither of these dates would be formatted as "9999," since even Y2K-susceptible applications use a six-digit date field (00/00/00) to represent the date. April 4, 1999, would be formatted as "04/04/99" and Sept. 9, 1999, as "09/09/99. "The "9999" string was selected as a nonsense or end-of-process date because it would not occur in normal operations using dates recognizable to humans."[112]

6. October 1, 1999. On this date, the federal government will begin its fiscal year.

7. January 1, 2000. THE DATE!

8. January 4, 2000. On this date, the first business day of the new year begins.

9. February 29, 2000. On this date, leap day occurs. Many assume that every fourth year is a leap year. But every fourth turn-of-the-century is a leap year, too. 1900 was not a leap year; but the year 2000 is. If the computer does not recognize February 29, errors will result. But also

some programmers did not know this, and their programs will neglect this date, increasing the problems.[113]

10. December 31, 2000. Since some computers don't know the year 2000 is a leap year, they will be confused when they reach this date, the seemingly impossible 366th day of the year.

Watch out for other events:

• If you hear any serious discussion of a "bank holiday" longer than December 25, 1999 to January 3rd, 2000. (We have heard discussion of a one–week bank holiday to help at the banks finalize all necessary Y2K preparations. The closing of all the banks during this one week does not trouble us in the least. In fact, we see this as a good thing. If it gets any longer than that, we would say it might be wise to consider some good old-fashioned PANIC! However, the likelihood of the banking industry having Y2K problems is almost non-existent.)

• If you hear any serious discussion of serious banking restrictions such as on loans or how much actual cash (remember, we are referring to those green printed pieces of paper that come out of ATM machines) you can withdraw from the bank on any given day, week, or month.

• If you are hearing nationally any serious discussion of a forced shutting down of all non-essential businesses and factories for the purpose of allowing more resources to be available for essential services. We would not view it as a major problem if various individual companies consider *voluntarily* shutting down for a short time due to Y2K related issues.

• If you hear any serious discussion of curfews, or restricting freedom of travel or movement (At first start of real talk by government officials to restrict freedom to travel, you better get out of the city as fast as you practically are able to and start preparing for the worst.)

• If you hear any serious discussion of rationing laws

- If you hear any serious discussion of the President issuing "Executive Orders" which is fancy, governmental double-speak for "martial-law" and "government takes over everything."

- If you hear any serious political or economic collapse in foreign countries.

If you start hearing any of these things, you will want to take out that "Plan B" list we keep talking about, and you want to start thinking about which of these things you need to do-NOW! If you start hearing of numerous things from our "finger-in-the-wind" check list, or similar types of things, you want to SERIOUSLY start implementing as many "Plan-B" check list items as you practically can. Again, we want to emphasize, this is ONLY if things start developing to look like we will be entering the "worst case scenario" effects of Y2K.

What we want to emphasize here is that we are basically more optimistic than most other authors writing about Y2K, as to what all will occur as a result of the whole Y2K situation. That does not imply that we do not think there will be any serious problems. There will be some serious problems. But we do not think it will be "the end of the world-as-we-know-it." We will be the first to admit that it is possible we can be wrong as to exactly how bad it might be. No one has the knowledge or skill to be able to predict exactly what will and will not be able to be fixed between the time we are writing this and January 1, 2000. Nobody can predict just how the governments will react to all this. More importantly, NOBODY can predict just how the general public will react as to all this. That is actually probably the greatest variable and wild card to all this. And that is the very reason why you need to keep your finger in the wind; right now! (Yes, we know, the grammar is wrong on that. We done that on purpose to get yous attention. So get yous finger in the air, start doing it right now, and keep it there-until January, 2000! Your spouse can help you if your arm gets tired!)

These are just some of the things to be watching for. We are sure there can be others, as we know you understand there can be.

Another thing of great importance is where to be watching for good sources of information on various key topics such as the computer industry, manufacturing, the banking industry, Fortune 500 companies, and international news. What we are going to do next is to give you some examples. And we know that not all of these will pan out to be good sources in the end. But it will give you a good idea of some of the places we will be looking. Nationally recognized business magazines and newspapers such as the *Wall Street Journal, Fortune, Forbes,* or *Business Week.*

Computer magazines written for computer specialists, or better yet, written for the computer industry. One of our favorites is a must-read type of weekly newspaper for the computer industry called *Computer Reseller News*, probably the single most important and comprehensive publication about what exactly is going on in the computer industry for both hardware and software. Another similar news weekly is *PC Week*. Both can be found in the periodical section of many libraries.

The Internet can be an excellent source of up to the minute information but it is difficult to weed out the wacko stuff. Often check out our web site

www.homecomputermarket.com For links to essential Y2K periodicals, information and updates.

Some of the web sites we found most helpful:

www.zdnet.com/zdy2k/ This is a great site for good news about Y2K.

www.y2ktimebomb.com

www.cfcministry.org This is Larry Burkett's web site.

www.cbn.org This is Pat Robertson's Christian Broadcasting Network web site.

www.y2kwomen.com This is a site of interest to women, particularly homemakers.

Signs we have been seeing!

While we have been researching this Handbook, we have found many ENCOURAGING signs. We felt we must include these items.

As Y2K awareness grows, resources will concentrate to solve the problem. Listen to what this Y2K writer says: "As the growing vision of the century-end "concentrates the mind" with escalating intensity, and the realization that "sound management" and "best practices" in

Q4 1998 means triage, contingency planning and confessions of the need to deal with potential system failure, and systemic failure, some of the leaders of the Year 2000 "community" have altered their message. That altered message—that "global meltdown" is unlikely, but that widespread inconvenience of uncertain duration IS—has not gone unnoticed inside the Washington Beltway."[114] Another programmer makes this point to consider: "We can accomplish a great deal by January 1, 2000. During World War II, in spite of our limited technological capabilities, we produced a Liberty ship every four days. We turned auto plants into rapid production lines for warplanes. We mobilized trains, planes, and automobiles—and factories, cranes, and dockyards—with unwavering focus and unprecedented logistical prowess."[115]

Even the cost of Y2K conversion is going down because new fixes and automated conversion techniques are being developed constantly. In 1997, estimates were running at about $3 per line of code. But a new fix from a company called Millennium Solution has reduced costs to "about 14 cents per line." Some solutions are able to run on a desktop PC, "so that the Y2K work was offloaded from the mainframe and didn't interrupt ongoing processes."[116] This saves the company from sending all work offsite. Another solution has "cut the time to make Year 2000 fixes on some software by 80% or more.... A consultant making Year 2000 fixes for Citibank, says [this] solution has allowed him to test and fix in one day a system that had 100,000 lines of code, compared with 30 days for another tool."[117]

Here are some press releases that clearly indicate that corporate America is hard at work fixing the Y2K computer problem. The fact that they are experiencing Year 2000-related computer errors now is a good thing. The more they encounter now, the more they will get them fixed before the deadline of January 1, 2000.

Full-Fledged Year 2000 Action Plans are Underway, Though Errors and Missed Deadlines Persist Companies Grow More Confident About Year 2000 Staffing and Spending, Survey Shows

Foreseeing Challenges in 1999, Corporate America Sharpens Year 2000 Testing Tools and Expands Contingency Planning

New York, NY, October 27, 1998—/Y2K WIRE/—"With 14 months until the new millennium, a new survey shows that America's largest corporations are increasingly experiencing Year 2000-related computer errors and missing deadlines in their Year 2000 plans.

"Yet while firms expect significant new challenges in 1999, 98 percent have launched full-fledged Year 2000 strategies which have helped them assume greater control over spending and staffing concerns. The tracking poll is one of the longest-running surveys to systematically monitor Year 2000 readiness. Covering IT directors and managers of 114 major corporations in 12 industrial sectors and 13 federal, state and local government agencies, it is conducted by Cap Gemini America LLC, a market leader in Year 2000 services.

"Increased contingency planning reflects a growing sense of realism about the magnitude of the Year 2000 challenge," said Jim Woodward, senior vice president of Cap Gemini America and head of its TransMillennium Services group. "Comprehensive testing, together with thoughtful contingency planning, will help protect organizations from major Year 2000 disruptions."

"Companies are adopting more rigorous testing protocols. The percentage of respondents who have established a process to check the quality of renovated code prior to testing increased from eight percent in April to 16 percent. And while respondents in July expected to use end-to-end testing methods for 20 percent of their applications, that figure has now risen to 30 percent.

"Nine of ten companies are now developing contingency plans to avoid Year 2000 computer-related failures. Sixty percent of respondents' contingency planning focuses on preventing a Year 2000 disruption and 30 percent concentrates on developing strategies should a Year 2000 failure occur.

"The Cap Gemini America survey again ranked 12 industrial categories and the public sector in respect to Year 2000 preparedness. Similar to the August results, the software, financial services, and computer sectors scored highest. The least prepared categories, listed by declining level of readiness, are pharmaceuticals and distribution (tie), health and transportation (tie), and public sector. Cap Gemini America is a member of the Cap Gemini Group, one of the largest computer services and business consultancy companies in the world.

Year 2000 Preparedness by Sector October 1998

Ranking	Category	Score	Number of Respondents
1.	*Software*	*88*	*10*
2.	*Financial Services*	*86*	*14*
3.	*Computers*	*84.5*	*8*
4.	*Manufacturing*	*83.5*	*10*
5.	*Telecommunications*	*82.5*	*10*
6.	*Aerospace (tie)*	*81*	*10*
6.	*Oil & Gas (tie)*	*81*	*8*
8.	*Utility*	*80.5*	*8*
9.	*Pharmaceutical (tie)*	*80*	*11*
9.	*Distribution (tie)*	*80*	*8*
11.	*Transportation (tie)*	*79*	*10*
11.	*Health (tie)*	*79*	*9*
13.	*Government*	*65*	*13*

Preparedness rankings are based on a combination of project progress and risk failure factors for each sector, based on survey responses." [118]

Here are other little press blurbs indicating a massive mobilization of manpower, resources and brainpower to develop software and tools to fix the Y2K computer glitch. (Now I know most of the computer–ese language will not translate (It doesn't to me either!) but you will get the basic idea.)

NEW YORK, November 23, 1998—/Y2K WIRE/—WOW! Microsoft® Access users from all over the USA, Canada, Europe, Australia, the Middle East, and even the Far East have been in touch with Serious Software, Ltd. recently. Just a couple of weeks ago, the New York software firm released ACC-FIX 2000TM, a new remediation tool that actually automates the process of fixing Year 2000 problems in Microsoft® Access desktop databases. And now it seems as if the entire Access user world has started knocking on Serious Software's door, along with Computerworld, CNN's Financial Network, Asia Pulse, Access World News, Canada Newswire, Access Advisor, and United Press International.

Houston, TX, November 16, 1998—/Y2K WIRE/—BindView Development Corporation, a leading supplier of systems management software, today introduced significant technology enhancements to its BindView EMS/NETinventory software for desktop inventory management and Year 2000 compliance assessment.

MORE HEADLINES FROM OTHER CAP GEMINI PRESS RELEASES:

October 14, 1998 - *Princeton Softech Unveils Year 2000 Testing Solution for Oracle Databases Move for Servers Improves Test Data Quality and Speeds Testing Process*

October 13, 1998 - *Millenia III Partners with Halliwell Engineering to Address Year 2000 Issues in Building Infrastructure Systems*

October 12, 1998 - *Princeton Softech Guides Companies Toward Year 2000 Compliance; Speakers Dispel Year 2000 Testing Myths at Global IT Forums*

October 6, 1998 - *Leading Year 2000 Tool Vendors Join Forces To Bolster Year 2000 Testing Integrity*

October 5, 1998 -*WRQ Express 2000 Software Manager Eradicates PC Millennium Bug Now; And Pays Back the Investment by Managing Other Issues Beyond 2000*

August 31, 1998 -*Scanalyzer Year 2000 Solution Version 2.50 improves Y2K Project Technology*

August 26, 1998 -*Novell Partners with Greenwich Mean Time to Speed Up Year 2000 Problem Resolution*

August 19, 1998 -*Princeton Softech Solves Major Problems in Aging DB2 Data Relational Tools*

July 27, 1998 -*Princeton Softech Markets Testing Boot Camp 2000*

July 24, 1998 -*Data Integrity Announces Year 2000 Audit/Pre-Test Tool*

July 21, 1998 -*Princeton Softech Year 2000 Productivity Tools Pay Big Dividends*

July 14, 1998 -*Princeton Softech's HourGlass 2000 Speeds Readiness for Year 2000*

May 12, 1998 - *Experts Write New Y2K Problem-Solving Guide for the Business World*

May 6, 1998 - *Y2K ALLIANCE OFFERS INSIGHTS INTO TURNING CRISIS INTO COMPETITIVE OPPORTUNITY*

And many more...

SECTION III

WHAT TO DO

CHAPTER 7

WHAT SHOULD CHRISTIANS DO?

As a person examines the whole Y2K issue, the biggest question that needs to be asked is, "What should *my* response be to all this?" Or put another way, "So what am I going to do to prepare for January 1, 2000?" Whether people ask themselves these questions consciously or unconsciously, they will be answering these questions either through their action or through their inaction.

If a person considers himself (or herself, of course) a "Christian", then the question must be asked from a different perceptive, "*As a Christian*, what should I do in light of the Y2K issue?" From this viewpoint, it opens up a whole different angle from which one must do all planning and preparation. If you are not really a Christian, there is not much point in trying to prepare for Y2K from a Christian perspective.

But before we can look at the whole Y2K issue from the Christian perspective, we must *analyze* one other critical issue in *significant detail*.

How *do* you *know* you are a Christian?

We are going to digress from the main subject here for a while. Whereas Y2K has the potential to be one of the greatest crises many of us will face in our lifetime, a far greater event awaits each and every one of us. Unlike Y2K, this other event has a 100% degree of certainty: Facing the Judgment Seat of Christ.

Whether or not Y2K causes major problems, and even if you manage to get through Y2K with little or no difficulty, God's Judgment will still have to be faced. This is a FAR more serious event. Y2K could possibly affect you for a lifetime, God's Judgment will, with all certainty, affect you for all eternity. And forever is a *long* time.

But what plans have you made for eternity?

We often make plans that we believe will affect us for the rest of our lives. Post high school education, marriage, career, children, etc. But with this question asked, we wish to examine the answers a little closer.

We realize that most the people reading this Handbook would label themselves as "Christian." We understand this. We also know that there will be some people reading this Handbook who would not consider themselves "Christian." A person would think it goes without saying "This section must be aimed at those who would not consider themselves Christians and have not yet made their plans for eternity." This is not necessarily true. Sure, it is meant for them too. But we actually want to address those of you who do consider yourselves Christian. We want to ask you to ask yourself the question "How do *I* know I am going to spend eternity in Heaven?" Or put another way, "How do I *know* I am a Christian?"

Most of you know of at least one person who would claim to be a Christian, but you believe there could be no possible way they could be a Christian. Either their actions or what they claim they believe are totally inconsistent with what the Bible teaches.

(Authors' note: This is not to imply that it is up to us to judge who is a Christian and who is not. But the Bible does give some very specific guidelines on what characterizes somebody as a Christian or not a Christian. This is also not to say we are saved by our own actions. But our actions do reflect what we truly believe. We will address this issue shortly.)

For example, there are many people who would like to make the claim they are a Christian, but they reject many of the basic fundamentals that the Bible teaches as part of the core of being a Christian. They determine what is right and wrong based on their thoughts, circumstances, feelings, or personal interpretation of what they want to believe the Bible says.

This is another issue we will get back to in a few pages.

Satan's strategy

Examine the issue of salvation from Satan's perspective. His ultimate goal is to keep as many people out of heaven as possible, thereby condemning them to an eternity in hell, with him. Look at his prime strategies. If Satan can get someone to believe God does not exist, his job is basically done. He has achieved his goal.

G. K. Chesterton put it this way, saying that those who won't believe in God will believe in anything. The logical extension of this is the current fascination with evolution and aliens from other planets. After all, if one could prove evolution, then one could say man could have come into existence without "god." The concept of the existence of aliens pushes this thinking one step further. If aliens exist, the (wrong) logical conclusion is that evolution has successfully produced intelligent life on other planets. This (wrongly) validates man's ability to come into existence without the need for a "god" to exist. The whole pursuit of evolution and alien existence is basically an intellectual attempt by people to try prove to themselves and others the absence of a need for "god" to exist in order to explain man's existence on Earth.

Why do people want to try to believe, through mental gymnastics, that we evolved "by chance" and that "god" does not exist? Because they do not want to have a "god" they have to be accountable to. They want to be mentally free to set their own standards and philosophies that allow the morality they desire to live by. This same thinking also is one of the primary reasons why many Americans choose the religion they want to believe. They find a religion that conforms to what they WANT to believe and justifies their choice of morality. They do this, instead of being willing to conform to the actual truth. We discuss this idea more in a few pages.

Satan, The Great Counterfeiter

Satan is the Great Counterfeiter and the Great Deceiver. For every Biblical truth, he creates many counterfeits or deceptions of that truth, with the goal to mislead, deceive, and hide the real truth from man. People have to stop envisioning Satan as a hideous, scary, Hollywood-movie-style persona. They have to start seeing him instead as the subtle, slick, con-artist that he really is. ALWAYS keep in mind that the Bible never depicts Satan like most people envision Satan to be. The Bible does not portray Satan as a wolf, like people do; the Bible portrays Satan as a wolf *in sheep's clothing.* There is an incredibly significant distinction between the two. The Bible NEVER portrays Satan the way Hollywood does in all its horror movies. Quite the opposite! The Bible calls Satan an Angel of Light.

Until we see and fully understand Satan as a deceiver and counterfeiter of the God's Biblical truths, we will never be able to see the falsehood in Satan's lies. If we cannot see the falsehoods, then we cannot protect ourselves from them. And just like an illness in our own body, if we are ourselves deceived, not only can we not protect others, but we will also be guilty of spreading Satan's lies and deceptions ourselves!

There are many places where the Bible warns of Satan's deceptions. Jesus warned of the many people who would, due to their *own* deception, attempt to deceive others. In Matthew 24:11 Jesus says "And many false prophets shall rise, and shall deceive many." Mark says in verse 13:5 "And Jesus answering them began to say, Take heed lest any man deceive you."

If someone does not think they are sick, they have no need for a doctor. Likewise, before someone can be "saved" they first must realize they are lost. So as long as Satan can keep people deceived into thinking they are on their way to heaven when they die, his deception has been successful.

There are no neutral powers. In reality there are only two powers, one being evil and one being good. And most often the evil power comes NOT as dark, terrifying, or wicked; but rather disguised— NOT wanting to reveal its true self, NOT wanting to reveal its true intentions or motives. The classic case is the "wolf in sheep's clothing."

A mugger makes no attempt to hide his intentions. He wants what rightfully belongs to you and he wants it now. No options, no discussion, no waiting, just "Hand it over." And he gives the threat of serious physical injury if you do not comply. On the other hand, a burglar would prefer to slip into your home unnoticed. His motives are the same. Just his method is different. He wants anything of value that is yours. And of course he takes it without your permission or approval.

And then there is the con artist. Bottom line, you hand him your money and he makes you feel good about it! This is exactly what Satan tries to do with people's souls.

Yes, certainly there is an obvious side to evil—crime, violence, and hate. If someone comes walking down the street towards you, and they look like a crook, and they are carrying a gun, you are able to put two plus two together and decide to stay as far away as possible from this person. Thus you avoid any of many possible disasters.

But the not so obvious side to evil is even more dangerous. Does the con artist use a gun to steal your money? Does he make his intentions known? When Dan was in the rare coin business, he had a great deal of contact with very many people who had been swindled out of their money by crooked coin dealers. Never did these victims describe the culprit as seeming to be evil, terrifying, wicked, or someone that a person would want to avoid. Usually, it was just the opposite. At first many of these victims refused to even believe they had been deceived and swindled. This is the ultimate example of the deceptive side to evil: when those trapped within it deny they are in evil's grasp.

We personally knew one of those people that turned out to be a con artist. By the time his scam fell apart, he had cheated people out of tens of millions of dollars. Did he seem evil to us? No. Did he seem like someone that would eventually be convicted of numerous crimes? Not at first. But Dan can still clearly remember the day that same man told Dan an absolute lie while giving Dan the friendliest, most cheerful smile he could. After enough lies, half-truths, and attempted deceptions, Dan realized that it would be best to have nothing to do with him.

Will Satan turn you into a con artist?

To illustrate this point, we want you to think of the term "typhoid Mary." She went around infecting other people with typhoid, not because she was intending to, but because she had no idea she was doing it. Since she "seemed" to be in perfect health, the last thing she thought she needed was a physician. And that is exactly what Satan's ultimate goal is. Not only to prevent a person from getting to Heaven, but also to get that person actually going around spreading a message that would cause others to end up in Hell also.

Satan's other tactics

The wicked tend to know where they stand spiritually and that they are at risk of eternal condemnation. Where Satan does some of his greatest works are not in the area of denial or rejection of God. Satan's greatest efforts are in the area of WHO or WHAT God is, and what is required to have eternal life in heaven. If Satan can *convince* somebody they have already found the way to 'eternity in paradise', they will stop looking. Why keep looking for what you think you have already found?

It does not take much investigation to realize there are about as many different forms of 'god' and as many different beliefs of how to get to 'heaven' as there are different cultures. There are many that would say there are MANY different ways to 'heaven'.

Contradictions are one of the greatest problems with many of these different beliefs. One belief says *this* must be done to reach 'heaven', The other says *that* must be done to get to 'heaven'. Unfortunately *this* and *that* are often in direct contradiction or opposition to each other. So consequently, the 'god' they want, turns out to be in contradiction with himself; he is inconsistent, irrational, and illogical.

God is a perfect God. A God of pure truth, justice, and logic. God is not inconsistent with himself. He would not give a way to heaven to one group of people, which is in direct contradiction to how He would direct a different group of people.

Another side of the "many different ways" belief is they want to believe in a 'god' which is not so narrow minded as to offer only one way to get into heaven. They want a flexible 'god' who is compassionate and understanding of different cultures and 'religions', tolerant of those poor savages who were never taught right from wrong.

And at the other end of the spectrum, this "god" is tolerant of good, moral people who strictly adhere to the teachings of their own religion. Their 'god' could never be limited to only "the narrow-minded legalism of those right-wing fundamentalist wacko kooks" beliefs of how to get to heaven.

Their 'god' could never send good, moral people to an eternity in hell.

Of course their 'god' never could do these things! Because, unfortunately, their 'god' does not exist, except for existing as a creation of their own mind. Instead of trying to seek after the true God, the Lord Jesus Christ, come as God in the flesh; instead of trying to learn what Jesus Christ taught, they wish to shape and mold a 'god' in their own mind. To form a 'god' that fits their expectations and desires of what they want their 'god' to be like.

The Bible calls this "idolatry." No, these people are not carving out statues of stone or wood. But they are literally carving and molding their 'god' mentally. Carving away aspects of God they do not want to accept. Adding to their 'god' attributes they wish and desire their 'god' to have. Mentally concluding that 'god' must be the way they want to believe 'god' is. But unfortunately their 'god' only exists in their imagination.

Obviously, there are many reasons why people choose to form a 'god' of their own making. We first heard this next expression said by someone else, and since then we have observed this truth all too often: People tend to believe what they believe in order to justify their own personal morality. When confronted with the truths of the Bible, they tend to dismiss many of these teachings in one way or other.

Even within what is broadly considered as "Christianity", there is a very wide spectrum of opinions and views. We refer to this as "Burger King Christianity: Have it your way!" Whereas this works fine for restaurants and menus, many want to apply it to "religion." But it does not work that way. An example of this is the classic response many have of wanting their own personal interpretation of the Bible. "You have your interpretation and I have mine" But sadly, the Bible does not open itself to "personal interpretations." Yes, a particular verse may have many different applications, but that is only one accurate interpretation. Instead of accepting Jesus Christ's teaching, many people want to twist Jesus' words into what they want it to mean.

One reason why people choose to form a 'god' of their own making is because they don't want to do what they think they might have to do. So to 'justify' their unwillingness to do the things they know they might have to do if they accepted the Bible fully, they reject the Bible instead. And, of course, they reject Jesus Christ as Lord and Savior. Or they try to re–create Him to suit their own needs.

Another mistaken belief is that a person has to stop doing all those wrong things and bad habits before becoming a Christian. This misconception is analogous to thinking a person needs to get cleaned up before they can take a bath. Becoming a Christian and being forgiven is not dependent on our becoming perfect first. Mt 18:11 says "For the Son of man is come to save that which was lost."

The differences between Biblical Christianity and all other religions

There are two key factors that separate all other religions and true Christianity. Both of these factors are related to each other.

The first factor is sin. Most organized religions basically recognized sin, its destructiveness, and how sin keeps people out of 'heaven' Sin is what separates man from a perfect, Holy God. It is the way each religion deals with sin that is the critical difference. All religions (and those who in error refer to themselves as 'Christian') deal with the issue of sin in a combination of two basic ways:

1) Strive to commit less sin

2) Attempt to balance the penalty of sin with good works and doing the "right" things.

Unfortunately, if the Bible is to be believed, neither of these things will save a person from an eternity in hell. Here is just a little of what the Bible says about sin itself. Romans 3: 23 "For all have sinned, and come short of the glory of God;" and Roman 6:23 "For the wages of sin is death."

The challenge between sin and right-eousness

Here are some verses about what the Bible says regarding trying to be good enough to meet God's standards:

Romans 3:10-12 "As it is written, There is none righteous, no, not one. There is none that understandeth; there is none that seeketh after God. They are all gone out of the way, they are together become unprofitable; there is none that doeth good, no, not one."

Romans 10: 3 "For they being ignorant of God's righteousness, and going about to establish their own righteousness, have not submitted themselves unto the righteousness of God."

The author of Romans makes the following conclusion through the inspiration of the Holy Spirit. He sums it all up by stating that it is IMPOSSIBLE for a person to be justified (saved) through his or her own actions. Romans 3:20 is the author's summary: *Therefore by the deeds of the law there shall no flesh be justified in his sight: for by the law is the knowledge of sin.*" The rules or 'laws' of the Bible are not designed to set a standard to try follow for entrance to Heaven. The primary reason God gave us the law is to show man (and woman) how wicked we truly are and how far we fall from God's standard of perfection. The best analogy we have heard is this. God's law is like a mirror. You cannot use it to clean yourself. It only serves to show you how dirty you are. Of course, the secondary purpose of the law is also to make life and interaction between people more civilized.

It is not a question of "how much sin to too much?" According to the Bible, ANY sin is too much. James makes this very clear and very specific, for those of you who think you can be good enough to make it into Heaven based on living a good, moral, and/or honest and upright life. He writes in James 2:10 "For whosoever shall keep the whole law, and yet offend in one point, he is *guilty of all.*"

"Justification" through Works

All this brings us to the second key factor which separates all other religions and true Christianity. In other religions, the key to reaching heaven is based on what the individual does in their own life to earn it, justify it, be worthy of it, or however else it is put. But no matter how it is put, each religion's form of 'salvation' is based on what the person does during their life.

For many religions, (including many who think they are 'Christian') at the point of 'judgment' after death, all one's actions in life are weighed like a balance between the good and the evil. In fact, the Bible even comments on the issue of trying to achieve righteousness based on one's own efforts or deeds. In Romans 9:32,33 it specifically answers the question of why true righteousness cannot be reached though one's own actions. "But Israel, which followed after the law of righteousness, hath not attained to the law of righteousness. Wherefore? Because they sought it not by faith, but as it were by the works of the law. For they stumbled at that stumblingstone"

It is not an issue of how good you have to be or how much good you need to do. According to the Bible, it is impossible to be good enough to earn or deserve the right to enter the most Holy presence of the Creator of the Universe.

With Christianity, salvation is not based on what we do to earn salvation or do to deserve it. Salvation is based on what Jesus Christ did on the cross at Calvary. Salvation is a free gift. And as a free gift, there is nothing that we, with our own power or strength, can do to merit or deserve it. If we could do something to deserve salvation, then it would no longer be free. We would have earned it through what we did, thus making Christianity no different than any other religion. But this is not the case. Salvation is a free gift offered to all that would receive it.

Is it possible that you can still end up in Hell, even if you "believe in Jesus"?

With all these methods which Satan uses to trick and deceive people, there is one deception above all else, which is the most subtle and dangerous of all: the "unsaved believer."

For many people who call themselves Christian, the answer to the earlier question ("How do you know you are going to spend eternity in Heaven?" Or, "How do you know you are a Christian?") focuses on the most well known verse in the Bible, John 3:16. *"For God so loved the world, that he gave his only begotten Son, that whosoever believeth in him should not perish, but have everlasting life."* And this is the foundation for their claim of Christianity and claim of eternal salvation. They 'believe' in Jesus Christ so therefore they will "not perish, but have everlasting life."

This 'belief' in Jesus Christ for salvation needs to be looked at in the light of another verse in the Bible, which examines the issue a little closer. James 2:19 states "Thou believest that there is one God; thou doest well: the devils also believe, and tremble." No one would argue "Satan does not believe in Jesus Christ." Of course Satan believes in Jesus Christ! Does that save Satan? Obviously not.

If we really understand what is being said here, we see that by itself, basic intellectual acknowledgment of the historic fact that Christ died on the cross for the sins of the world will not get us into Heaven. There is more to salvation than this. We will get to what exactly this is in a moment, but before we get to it, we need to cover the "belief" issue in better detail.

The Bible clearly states things like Matthew 7:13 "Enter ye in at the strait gate: for wide is the gate, and broad is the way, that leadeth to destruction, and many there be which go in thereat." More specifically to this point is Matthew 7:21-23 "Not every one that saith unto me, Lord, Lord, shall enter into the kingdom of heaven; but he that doeth the will of my Father which is in heaven. Many will say to me in that day, Lord, Lord, have we not prophesied in thy name? And in thy name have cast out devils? And in thy name done many wonderful works? And then will I profess unto them, I never knew you: depart from me, ye that work iniquity."

Just who are these people that Jesus is referring to? Or maybe a better way to pose the question is exactly what is it that separates these people from God? Answer: **SIN**.

Jesus is referring here to people who recognized Jesus Christ truly existed and was not a myth or fable. These people did good works in Jesus' name, to honor Him. But still they died in a state of separation from Him. Remember that Satan ultimately does not care WHAT they believe, even if it is PART of the truth. As long as they do not know the full truth and the true way to be completely forgiven from sin, Satan can use any portion of the truth will work as long as it keeps these people in a state of unforgiveness, and eventually going to hell.

The significant question is why? Why will Jesus say to these people, at Judgment Day, "I never knew you: depart from me, ye that work iniquity."

A favorite evangelist of ours, Major Ian Thomas, explains it this way. He refers to these people as "unsaved believers." Unsaved believers are people who do believe Jesus Christ was crucified on the cross for the sins of the world. They do intellectually acknowledge this as a historical fact. But they, personally, are not forgiven of their sins. Certainly belief in Jesus Christ is a component of salvation, but not the sole ingredient.

To really understand why Christ will say to these people on judgment day, "I never knew you: depart from me, ye that work iniquity." and to understand just why these people died with their sins unforgiven, we need to take a closer look at the issue of righteousness and forgiveness and how it is related to salvation. All of which also brings up the issue:

How *to know* you are a Christian!

Now for those of you who have full assurance of your own salvation, please do not become insulted or offended by this section. Use this as an opportunity to really understand what you believe, and to be able to share it in a logical and accurate way with others.

For those of you who are not 100% sure of your salvation, use this as a challenge to examine what it is you believe and what is the basis of why you believe it.

We already wrote this once, but it is important enough to say it again, because it brings the next section into proper perspective. If someone does not think they are sick, they have no need for a doctor. Likewise, before someone can be "saved" they first must realize they are lost. So as long as Satan can keep people deceived into thinking they are on their way to heaven when they die, his deception have been successful.

Repentance

One of the most common themes throughout the entire Bible is the call for us to repent. And this is the very first step in salvation. Someone who thinks they are healthy will think they have no need to be healed, and no need of a doctor. Likewise, until we see our sin and our need for a Savior, we will not begin to take the steps we need to tale towards salvation.

Once we see past Satan's first deception (that we have no need for salvation or that we already think we are going to 'heaven') and we recognize our condition as a sinner, we need to repent.

These are just some of the verses in the Bible that refer to repenting.

Mark 2:17 When Jesus heard it, he saith unto them, They that are whole have no need of the physician, but they that are sick: I came not to call the righteous, but sinners to repentance.

Eze 18:30 Therefore I will judge you, O house of Israel, every one according to his ways, saith the Lord GOD. Repent, and turn yourselves from all your transgressions; so iniquity shall not be your ruin.

Mt 4:17 From that time Jesus began to preach, and to say, Repent: for the kingdom of heaven is at hand.

Mr 1:15 And saying, The time is fulfilled, and the kingdom of God is at hand: repent ye, and believe the gospel.

Lu 13:3 I tell you, Nay: but, except ye repent, ye shall all likewise perish.

Ac 2:38 Then Peter said unto them, Repent, and be baptized every one of you in the name of Jesus Christ for the remission of sins, and ye shall receive the gift of the Holy Ghost.

Ac 3:19 Repent ye therefore, and be converted, that your sins may be blotted out, when the times of refreshing shall come from the presence of the Lord;

Ac 17:30 And the times of this ignorance God winked at; but now commandeth all men every where to repent:

Lu 24:47 And that **repentance** and remission of sins should be preached in his name among all nations, beginning at Jerusalem.

Ac 5:31 Him hath God exalted with his right hand to be a Prince and a Savior, for to give **repentance** to Israel, and forgiveness of sins.

Ac 20:21 Testifying both to the Jews, and also to the Greeks, repentance toward God, and faith toward our Lord Jesus Christ.

2Pe 3:9 The Lord is not slack concerning his promise, as some men count slackness; but is long-suffering to us-ward, not willing that any should perish, but that all should come to repentance.

We think these verses speak for themselves pretty well and need no clarification.

As just mentioned, the first aspect of salvation is acknowledging our sin, the need to repent, and recognizing our actual need for for-

giveness. This brings us to the source of our forgiveness and therefore our Savior.

Jesus as Savior

Jesus is not *just* a great prophet. Jesus is not *just* sent from God.
Jesus *is* God. And part of salvation is believing and accepting this,
that Jesus is God actually come in the flesh.

John 11:25-27 "Jesus said unto her, I am the resurrection, and the
life: he that believeth in me, though he were dead, yet shall he live:
and whosoever liveth and believeth in me shall never die. Believest
thou this? She saith unto him, Yea, Lord: I believe that thou art the
Christ, the Son of God, which should come into the world."

There are a number of key parts of these verses. What exactly is
the specific thing that Jesus is asking Mary about in these verses?
Look at her response to Jesus' question for an insight to the critical
factor here. Obviously, she believed He existed, because he was
standing in front of him. So that is not the key. The key is that she
believed that Jesus was "the Christ" and "the Son of God," meaning
she believed He was the Messiah, the forgiver of sins, that He was the
forgiver of her sins. She believed that He was not just sent from God,
but that He was God, come to earth in the flesh.

Rarely has any religious leader ever claimed they were God, the
creator of the universe, as Jesus Christ did in so many verses through
out the Bible.

This brings us to another extremely significant difference between
Christianity and all other religions. The founders and leaders of every
other religion ever are all dead and buried, or will be dead and buried.
Jesus Christ and Jesus Christ alone is the only one to rise from the
dead. And one of the significances of His rising from the dead is that
all who put their faith solely in Him and in what He did on the cross,
will rise from the dead and join Him in Heaven.

John 14:6-11 "Jesus saith unto him, I am the way, the truth, and the
life: no man cometh unto the Father, but by me. If ye had known me,
ye should have known my Father also: and from henceforth ye know
him, and have seen him. Philip saith unto him, Lord, shew us the
Father, and it sufficeth us. Jesus saith unto him, have I been so long
time with you, and yet hast thou not known me, Philip? He that hath
seen me hath seen the Father; and how sayest thou then, Shew us the

Father? Believest thou not that I am in the Father, and the Father in me? The words that I speak unto you I speak not of myself: but the Father that dwelleth in me, he doeth the works. Believe me that I am in the Father and the Father in me: or else believe me for the very works' sake."

Uncertainty or Eternal Security?

There are many people who believe a person can lose their salvation if they are not 'good enough' or if they sin too much after they are saved. They try to use some Bible verses to make the point, and they try to use some analogies to illustrate the point. Unfortunately, the analogies tend not to be consistent with the basic themes of salvation.

If we really understand the issues of salvation, forgiveness, and righteousness, it puts to rest the question of "Can someone lose their salvation?"

If you understand that salvation is a result of forgiveness, and forgiveness is a result of becoming righteous, and that righteousness is not a result of what we did, but what Christ did, then the issue becomes clearer.

Look at the issue of losing salvation from a different perspective, the perspective God sees it from. When we are saved, we become dead to our old sinful nature. Our 'old man' is crucified with Christ, we are buried with Christ, we are raised from the dead with Christ, and we are "born again" with Christ. (We really do not like to use the words "born again" because this term has virtually lost its true, original meaning in the day and age we live in.)

Examine the issue of "Born Again" more closely. By being "born again" the Bible says that we are now born into the Kingdom of God and that we are now children of God. We ask you to think this through a bit. Is it possible to be 'unborn'? Or put another way, no matter what you do or say, no matter how hard you try, no matter how much you may hate your earthly birth parents, is it NOT possible for you to stop being the child (the son or daughter) of your parents. You just cannot.

You can change your name, you can change your citizenship, you can get adopted, a rare few would even try to medically change their gender, but not matter what you do, you will forever be the child of your parents. And likewise, those born into the kingdom of God are now His children, and thus they are sealed by the Holy Spirit and eternally secure in their salvation. How did they become God's children? Was it by what they did? No. It was not by what they did. And so we, through what WE do, can no more lose our salvation by what we do, then we can earn or gain our salvation by what WE do.

"Grace and Works are a Dangerous Mix"
or
"And What Discussion of Salvation Could be Complete Without Discussing Eternal Security"

We admit this next section will be very hard for some people to accept. And we understand this. But we ask those of you who disagree with us not to throw out what we are saying until you have compared your views with what the Bible says on the issue. Is your personal view of forgiveness and salvation consistent with what the Bible says about why Christ came and what was the purpose of His death on the cross?

If a person's understanding and belief of salvation includes the belief that one could lose one's salvation, then we believe they do not really accurately understand the concept of salvation. Or put another way, if you believe you can lose your salvation, you do not really understand what salvation really is. We will explain.

If a person is truly saved by grace and grace alone, then they cannot actually do anything within their own powers and abilities to become saved. It is either a free gift through Christ's finished work at the cross, or it is earned through one's own works and actions. Romans 11:6 says "And if by grace, then is it no more of works: otherwise grace is no more grace. But if it be of works, then is it no more grace: otherwise work is no more work."

If someone believes they are saved by grace, but they can lose their salvation, they are basically saying that once they become saved, they must "earn" the right to stay saved, through "their" actions. This means they are, after the point of salvation, no longer relying on what

Christ did on the cross. They are now ultimately relying on "their own" ability to retain their "good standing with God" through their own works.

"Probationary Salvation."

People who think they are able to lose their salvation have what we refer to as "Probationary Salvation." They do not believe in the concept of salvation through grace, as in a full pardon from Christ. They are out of "prison." And they can stay out of "prison," but only as long as they can uphold the terms of the probation. But, if "they" violate the terms of the probation, they are no longer free, but back in prison. Freedom is no longer based on an undeserved, unearned "presidential" (or Christ's) pardon. Instead, they are released from prison, but staying free is based on their own ability to stay within the "Law" and not violate it enough to lose their freedom.

If they actually can lose their salvation after being forgiven and saved, they are now in the position of ultimately being responsible for their own salvation. Christ has done his work on the cross, now it is their personal responsibility to maintain a level of righteousness that is acceptable to God for eventual admission to Heaven.

The more you look at the concept of losing one's salvation, the more you can see the blasphemy of it. When it comes to the final decision as to whether or not a person will be able to enter Heaven, this belief actually elevates the significance of that person's actions, after they are "saved", OVER the actions of Christ on the cross. The belief of being able to lose one's salvation means that Jesus Christ's shed blood on the cross was a start, but it was not sufficient by itself. Their own capability of staying free from sin is also required as the final work.

Look at the flip side of this same coin.

If a person truly believes they can lose their salvation, then one has to question if they are actually saved. (Now, we know we may be stepping on a few toes here. But this issue is too important to treat lightly. We apologize if we are coming across too strong but like the apostle Paul, the Lord compels us to preach the true gospel whatever the cost, at the risk of offending some.) If a person believes they can lose their salvation then, in the end, they are no longer actually believing in what Christ did, but they are ultimately putting their trust in their own ability to be righteous enough to enter Heaven. In the end, you are either 100% fully relying on what Jesus did for you at the cross, or you are also relying on your own righteousness to some degree. In which case, if a person is also relying on, in ANY way, to ANY degree on their own ability to refrain from sinning as ANY part of the terms of admission to Heaven, then they are placing their personal righteousness above what Jesus Christ did at the cross. They do no longer believe in salvation as a free gift, that Christ has done it all. They are now adding their efforts to the mix.

And is that not the exact same lie Satan is trying to put forth with nearly every other religion?!?

"Ask Jesus into your heart"

A variation of the great "unsaved believer" deception Satan has placed on Christianity today is the concept of salvation by "just ask Jesus into your heart." This concept is simply not found anywhere in the Bible! No where does the Bible tell us, ask us, or even suggest that we should "ask Jesus into our heart!"

There are many Biblical references to the heart in the New Testament. Verses such as:

Romans 10: 9-10 That if thou shalt confess with thy mouth the Lord Jesus, and shalt believe in thine heart that God hath raised him from the dead, thou shalt be saved. For with the heart man believeth unto righteousness; and with the mouth confession is made unto salvation.

Ephesians 3:17 "That Christ may dwell in your hearts by faith; that ye, being rooted and grounded in love."

What Satan has done is made an extremely deceptive, subtle counterfeit of the Biblical call to "believe in our heart." as referenced in Romans 10:9-10. Yes, Jesus DOES come into our heart, but this is a RESULT of our salvation. The act of "asking" Jesus to "come into our heart" is NOT the SOURCE or the method of our salvation! Nowhere does the idea of "asking Jesus into your heart" take into account the steps the Bible asks us to take. Asking Jesus into one's heart does not acknowledge our own sin, our own failure, our own inability to be righteous, or the sovereignty of Jesus Christ as God and Lord of our Life, and the only source of forgiveness through His death the cross for our sins. THESE are the key elements to salvation!

One of the passages of scripture where this misconception might come from is in Acts 8 where Philip is told by the Spirit to witness to the Ethiopian eunuch. Notice the following things. Philip never tells the eunuch to be a better person, to clean up his life and stop all his sinning. What Philip does is tell the eunuch about Christ's death on the cross as the ultimate sacrificial lamb, dying for our sins. As the eunuch hears all this, he realizes his own personal needs. Philip refers to "believing with all thine heart," but not here, or anywhere else in the Bible is there ANY suggestion ever made to "ask Jesus into your heart" or any similar concept.

Acts 8:27 "And he arose and went: and, behold, a man of Ethiopia, an eunuch of great authority under Candace queen of the Ethiopians, who had the charge of all her treasure, and had come to Jerusalem for to worship, was returning, and sitting in his chariot read Esaias the prophet. Then the Spirit said unto Philip, Go near, and join thyself to this chariot. And Philip ran thither to him, and heard him read the prophet Esaias, and said, Understandest thou what thou readest? And he said, How can I, except some man should guide me? And he desired Philip that he would come up and sit with him. The place of the scripture which he read was this, He was led as a sheep to the slaughter; and like a lamb dumb before his shearer, so opened he not his mouth: In his humiliation his judgment was taken away: and who shall declare his generation? For his life is taken from the earth. And the eunuch answered Philip, and said, I pray thee, of whom speaketh the prophet this? Of himself, or of some other man? Then Philip opened his mouth, and began at the same scripture, and preached unto him Jesus. And as they went on their way, they came unto a certain water: and the eunuch said, See, here is water; what doth hinder me to be baptized? And Philip said, If thou believest with all thine heart, thou mayest. And he answered and said, I believe that Jesus Christ is the Son of God. And he commanded the chariot to stand still: and they went down both into the water, both Philip and the eunuch; and he baptized him."

Yes, we need to "believe in our heart" before we can truly accept what Jesus Christ has done for us. Yes, Jesus Christ comes "into our heart" though the Holy Spirit as part of our salvation. But nowhere is anything said about "asking Jesus into your heart" found in the Bible. If you are trusting your salvation on a past request of asking Jesus into your heart, you may be heading towards the right direction (the need for forgiveness by Jesus Christ) but you are not there (salvation) yet.

We challenge EVERY one of you reading this to re-examine the basis of why you believe you might be going to heaven after you die. See if what you believe fits in line with the next two precepts as repeated over and over again in God's Word.

In Summary:

We are saved by admitting our sin (confessing and repenting), acknowledging our need for forgiveness (grace), THROUGH our faith (believing) in what Jesus Christ as God has done FOR us at the cross. We are NOT saved by what WE do, or by what we don't do ourselves.

We challenge you to take the time right now and re-evaluate your own faith. Have you:

1) Truly repented of your sins?

and

2) Acknowledged your total, absolute need for Christ's incontrovertible grace as the complete and sole source of your salvation, without relying on yourself and your own goodness for any part of your righteousness?

If you have not ever truly done this, RIGHT NOW would be the best time to do so in a prayer to God. We ask those of you who need to do this, to put down this Handbook right now and settle this issue with Jesus Christ.

The rest of the book can wait!

What's next?

Once you have received the Lord Jesus as your savior, your next step is to start reading your Bible every day. The books of John, Acts, and Romans are the best places to start. Secondly, find the best church you can and attend regularly. Refer to Chapter 16, *"Choosing the Right Church for 2000 and Beyond."*

Bill Gates and Y2K

A person can make all their Y2K plans and get through the year 2000 with no problems. But when that person dies, if they end up believing a lie, having been deceived by Satan, they will spend eternity in hell. We will put it another way. If a person had all the wealth and power of Bill Gates, if they were the wealthiest man in the world, what good would that do if they lost their soul and spent an eternity in hell?

This is the bottom line reason why we spent so much time on this issue. One soul is far more valuable than all of our successful Y2K planning. And if through this information one soul is saved, it is far better then all the other successful Y2K preparedness that comes out of this Handbook.

The Christian's response to Y2K
or
The Three R's of Y2K: Repentance, Redemption and Revival

(At this point we will be writing specifically as Christians, coming from a Christian perspective, and writing for Christians in terms of preparation priorities. However, this does NOT mean that our practical suggestions and recommended personal preparation ideas would not be beneficial to a non-Christians as well. The advice will still be sound. It is the motivation and priority levels that would be different for a non-Christian compared to a Christian.)

Let us outline the steps for the Three R's of Y2K.

Pray.

The first action a Christian needs to do is pray. And we do not mean pray Y2K won't happen. It will happen, and no amount of prayer is going to change that. What you need to pray about is what

YOU should be doing. As a Christian, ask God for wisdom as to what your response to all this should be. You need wisdom to make Godly decisions in four areas.

- What does God want you to do in your own personal preparedness? Are there any areas God is calling you to personal repentance? Does your church need to seek repentance?
- Are you at the church God wants you to be it?
- What does God want you to do to be a witness to those whom you know personally (friends, neighbors, co-workers, relative, etc.)?
- What does God want you to do to be a witness and servant to your community leaders?

Now, as we challenge you to seek after God for his wisdom, we realize He will lead different people in different ways. There will be some that will be led of God to move somewhere else, and there will be some that will be led to stay.

"Read your Bible, pray every day and you'll grow, grow, grow."

In addition to prayer, search the Scriptures diligently to seek God's heart and will in this matter. Since this is another topic with volumes of books on the subject, we will not go into any further detail. Besides, your pastor will probably preach about it this Sunday.

Obey God's will, not your will.

Be careful not to delude yourself. Make sure that whatever you believe God is calling you to do, you are doing it for Biblically sound reasons. This is a very important issue. If you do "all the right things" but you are not doing them in God's will, there will be consequences. The Bible refers to this as iniquity: doing what you think is right, instead of what God has called you to do.

There is a flip side to this as well. James 4:17 says "Therefore to him that knoweth to do good, and doeth it not, to him it is sin." When you are prompted by the Holy Spirit to do something, if you know it is God's will, then do it.

Seek Counsel

We would also advise you to seek good, Godly counsel from the authorities God has placed over you, before you make major decisions or purchases. Do not be too proud or arrogant to ask counsel from parents, in-laws, church leaders, and/or even employers. There is much that could be said about how God works through those he has placed over us, but this is not an issue we will be going into detail here.

Focus on the Three R's

Following these three steps will bring you closer to the true Three R's of Y2K: **Redemption, Repentance, and Revival.**

It is our hope and prayer that the material we have just covered will be clear enough to provide the necessary knowledge needed towards redemption (salvation) for any who read it.

As already covered, a part of redemption is acknowledging our need to repent from our sins. Although redemption forgives us of our sins, it does not stop us from sinning. That is our job, through the power of the Holy Spirit that lives within us. This is where both personal repentance and revival come in.

All of us already know some things that need to be eliminated from our lives. So this is a call to take the steps towards that end. Next, seek out an older or more mature Christian and have them look at your life to show you some of the "blind spots." Blind spots are areas you may not see, but others do see. Find out areas you may be making compromises that you have not seen or been considering. As you go about your day to day life, start with this next idea. As you make various choices and decisions each day, start choosing what Jesus would choose for you.

As we turn from our sins, and humbly turn to God in submission, focusing our lives on Him living in us, He will bring revival in our lives.

As revival breaks out in us, it will spread and break out in His Church. And this is what we need, first and foremost, as we go into the new millennium. As revival breaks out in the, then we will see great numbers of the lost coming to Christ.

The words of Zachariah are so appropriate to this time of uncertainty. In the end of the first chapter of Luke, John the Baptist was newly born and given his name. After God caused Zachariah to be mute during the entire pregnancy, God gave Zachariah his speech back again. His first words were a prophecy through the filling of the Holy Spirit. He spoke of why Christ was coming; "to give knowledge of salvation unto his people by the remission of their sins." But he called God's people to serve God, "without fear, in holiness and right-eousness before Him, all the days of our life."

Y2K should not be a time of fear and anxiety but a time of redemp-tion, repentance, and revival amongst God's people. So repent, get revival in your life. And look first to serving God and His kingdom, starting now and in the next millennium.

Are you where God wants you to be?

Start your Y2K preparations with your church. Get in the church where God wants you to be. (See Chapter 16.) If you are already in the church you belong in, what can you and your family do for the church? What can you and your family do to lift up the other mem-bers to greater spiritual maturity? What can you and your family do to help your church, as a body of believers, to reach out to your com-munity for Christ?

The answers to these questions will be different for everyone. Look first to Jesus Christ, the example of His disciples, and His words in the Bible.

Decide that you want to be a witness to those you know

This Handbook is not meant to be a theological textbook on the methodology of witnessing. What we desire to accomplish here is to *motivate* Christians to witness. We have listed a few ideas as sugges-tions to get you started. Just be creative and be open to the prompt-ings of the Holy Spirit. You could also ask other Christians for ideas and suggestions.

One of the most common times for people to turn to the Lord is in times of extreme need or crisis. This could be such a time. And as such, this could open up people you know to consider the Gospel for the first time ever.

Friends

Awareness of the Y2K problem isn't limited to computer experts, corporate leaders, and government officials. There is becoming a much wider awareness of it among the general public. Especially with more and more articles in mainstream publications such as USA Today, Time, Newsweek and such, along with occasional nightly news reports. Subsequently, this is something that more and more average citizens are examining. As they look into it, more people who are as "normal middle-class" as can be, are considering making significant plans to prepare. As January 1, 2000 draws closer, more and more of these "otherwise normal" people will start joining the ranks of what (in another time) would be thought of as "religious kooks."

We would like you to use Y2K as a great opportunity for you to witness to your friends. For some examples of things to do, look in the next **Neighborhood** section and Chapter 14, *"Your Church and Y2K"* and follow some of the same ideas.

Neighborhood

We recommend that one by one, you start inviting neighbors over for a dinner or some other social activity where you have the freedom and opportunity to talk to them in a private, relaxed environment. If dinner doesn't seem to work for you, try something different. Guys: go fishing, golfing, hunting, watch a sports event, or some other activity that works for you. Ladies: have her over for coffee, go on some walks together, take the kids to a park. Be creative.

Now, we realize it should go without saying that these would also be good things for witnessing. We realize this, and we are not trying to draw away from witnessing. But, remember this about witnessing. One of the examples set for us by Christ was to meet peoples greatest physical needs first, then look to meet their spiritual needs. Also check out Chapter 14 on church preparedness.

We want to make it clear that these are only some ideas and suggestions. We make no claim to know what will work best for you, because (obviously) you know you better than we know you. And you know your neighbors better then we know them. We also make no claim on being spiritual authorities on the subject of witnessing. In fact, we confess this is an area of weakness for us personally. You need to be open to two things at the same time. Be open and watching for signs and opportunities to bring up either Y2K or the Gospel with those around you. At the same time, be open to the leading and prompting of the Holy Spirit to what God would have you do. Especially be open to those promptings when you aren't necessarily consciously thinking specifically about what you should be doing. (We are aware that you will need to be thinking, every now and then, about things other than Y2K or witnessing!)

When you start to find a small number of neighbors who do express genuine concern over the potential prospects of Y2K, suggest to them some sort of get–together to discuss the issue. In many communities, groups are forming already to plan preparedness tactics. Here in our city the community college is already planning to be fully functional on New Year's Eve with the generators online to "be a resource to other governmental units and the general public."[119] In the city of Miami, Chuck Lanza, Director of the Miami-Dade County Emergency Management Group has proposed "that a reasonable community goal is for every household to be self-sufficient for 14 days."[120] These types of emergency preparedness recommendations are becoming more and more common throughout many communities.

Here are some tips to help your meetings be productive:

• At your "meeting" discuss potential problems and what each of you can do to prepare.

• Have printed material on hand that includes referrals to other resources and make other brochures, books, or informational packets available for purchase.

• Based on the magnitude of this issue, you will want to plan on meeting somewhat regularly, such as once a month.

• Try to raise awareness by sticking to a mainstream credible approach. Avoid specific predictions or dire warnings.

• As you get together, one of the many things you will need to address are what sort of things you can do collectively to prepare. There are different things that will be practical for different areas.

• Group together to buy preparedness items in bulk: Items such as food, commodities, generators, etc. are much cheaper when you order larger quantities.

• Organize a Year 2000 awareness group at church. Read Chapter 14, *"Your Church and Y2K."*

Here are some resources to help your neighborhood Y2K awareness and preparedness efforts:

• Contact community leaders to help start a community or neighborhood awareness and preparedness group. Contact Joseph Project 2000 for more information:

> Joseph Project 2000
> 6406 Bells Ferry Road
> Woodstock, GA 30189
> Phone: 678-445-5512 Fax: 678-445-5503
> E-mail: info@josephproject2000.org
> Web site: www.josephproject2000.org

• Rent a video about Y2K to watch with neighbors, church members, and friends. One good source is CBN. Pat Robertson's organization has prepared an excellent set of videos. Call 800-777-8398 to order *Y2K and the World* (aired July 10, 1998) and/or *Y2K and the Church* (aired August 7, 1998)

• The Cassandra Project: a grassroots group trying to organize communities to face Y2K problems.
Web site: www.millennia-bcs.com

As your group pulls together for Y2K you will naturally grow closer socially as well. The time will come where the discussion of preparing for a coming potential crisis will make for the opportunity to bring to their attention the *inescapable* coming of a far, far greater crisis they must eventually face: Judgment at the Throne of God. And then the question of what have they done to prepare for it?

CHAPTER 8

WHY PREPAREDNESS IS A GOOD THING OR THE REASONS FOR PREPAREDNESS

What is PREPAREDNESS?

In James Talmage Steven's book, *Making the Best of Basics: A Family Preparedness Handbook* (Which, by the way, you should purchase immediately. More on that later.), he defines Family Preparedness as...

"[The] practice of being aware and alert to the possible and probable reality of a disaster happening to you.

1. Preparedness is knowing how to:

- Ascertain potential disasters, whether natural, man-caused or personal, to which one's family is vulnerable and

- Eliminate or minimize risks- and therefore the negative effects- of any disasters to the degree possible.

2. In-home storage is how you can minimize the negative impact of the unexpected by having stored in your own home adequate resources of water, food, fuel, medical supplies and medications, clothing, money, transportation, bedding- anything you need to be self-sufficient during an emergency for an extended period of time."

We also need to make it very clear what preparedness is *not*. Preparedness is *not* "survivalism." Preparedness is not the mentality that the world as we know it is coming to an end. Preparedness is not joining an extremist group. Being prepared does not mean we have no faith in God to provide (Joseph prepared for seven years.) Preparedness is not a "wacko" thing to do. Preparedness is not planning for disaster (it is being ready for disaster, if it hits.)

Planning YOUR Y2K preparedness strategy

When you begin to "prepare to prepare", you need to understand the risk or likelihood of a specific disaster and the subsequent consequences of not being prepared. *The greater the chance of an event, the higher payback preparedness gives you.* In some areas, some types of disasters are more probable; others can be very unlikely. Mathematically speaking, there is a statistical possibility a meteor could hit you within the next year. However, the probability of it happening is almost non-existent. Therefore, no one prepares for the consequences of being hit by a meteor. Mathematically speaking, again, there is a significantly higher statistical possibility you could be involved in a traffic accident within the next year. Therefore, most people take not just one, but a significant number of steps to both reduce the probability of becoming involved in a traffic accident and to reduce the effects if they are actually involved in a traffic accident.

Examples:

To reduce the statistical probability of becoming involved in a traffic accident we take pre-cautionary measures, like fixing the brakes and making sure our cars are mechanically sound. At times of unusual weather conditions (heavy fog, icy roads, thick snow, excessive heat, etc.) we either reduce our normal driving speed or sometimes forgo driving altogether to eliminate the risk factor completely.

We also take additional measures to reduce the actual damages of potential accidents, in case they do occur. The first is one we do before we even get in the car. Nearly every one of us has auto insurance of some kind, to reduce the potential risks of damages to our finances. When we get in the car, we put our seat belts on.

So what do meteors and auto accidents have to do with the Y2K problem? We will answer that question after we discuss floods.

We have a bold confession to make. We do not have flood insurance. Not only that, but we refuse to get flood insurance as long as we live in this house. Now, we realize that some of you might feel we are just flirting with disaster. And if we were some of you (especially those of you who have gone through floods), we might feel the same way, "This guy is crazy!" Except you don't know what we know. If our house gets flooded, we will not need flood insurance. We will need an ark. We live on a hill, near the highest point of the county. So, although a risk exists for some other people in our county, it does not exist for us. Since there is virtually no probability of risk of flooding, we have no need of buying flood insurance. No probability of high flood consequences equals no risk, and therefore no flood insurance.

On the other hand, our home could burn down. Very low probability multiplied times very high consequences equals too much risk to ignore. So we do have fire insurance. Life insurance. Same scenario. Very low probability multiplied times very high consequences equals too much risk to ignore. So we do have life insurance. We don't have insurance because we think these things *will* happen. Or even because we think they *might* happen. We don't know exactly what the probabilities are for either, but the stakes are just too high to gamble with.

On the other hand, we know there is a school of thought that says "I don't need insurance, God will protect and provide." Yes he can, and yes he does-sometimes. But not every time. There is also a school of thought that says, "If we have a disaster, my Brothers and Sisters in the Lord will help me." But look at the other side of the coin, this is a two way street. YOU are also to be ready to help THEM if they have the need. Not only that, but also what happens when trouble hits everyone? When disaster hits one or two, the resources of the many can lighten the burdens of the few. But look at the incredible floods at North Dakota and Minnesota in the spring of 1997, virtually wiping out entire communities. Or the fires in Florida the in the summer of 1998. These major disasters wiped out everyone. Disasters have a tendency to wipe out both Christians and non-Christians alike. Faithful and trusting, or the unfaithful and untrusting; those who prepared in advance and who were insured were the best protected.

All this and we still refuse to buy flood insurance.

We wonder if we are covered for meteors!

Now, we realize you probably know where we are heading with all this. So are we done yet? No. We need one more example.

Hurricanes.

Sometimes a disaster gives you virtually no advance warning. Tornadoes. Auto accidents. House fires in the middle of the night. Compare that to hurricanes. Sometimes people have up to several days warnings before a hurricane hits. Enough to do some preparing. Board up the windows and bring things inside. Get emergency supplies (if any are left at the store). Fill up the car with gas, and get ready to evacuate when necessary. But in the end, in the big picture there isn't really much you can do to prevent major disaster.

What if you knew six months in advance that a hurricane was almost certain to strike somewhere very close to where you live? What might you do then? Same things? Board up the windows. Bring things inside. Get our emergency supplies. Fill up the car with gas, and get ready to evacuate when necessary? No? With enough advance warning, you could move those things valuable and sentimental to you to a safe location. You might even sell the whole house and move everything to a safe location. (Now let's not draw too much out of this example. we are NOT telling you to sell your homes and move to "safer ground." We will deal will this issue later. Nor is it appropriate here to have a discussion about the ethics of selling to someone a house which might very likely get wiped off its foundation. This is just an example for planning ahead. Nothing else.)

As we were saying before we so rudely interrupted ourselves, if you knew in advance that there was a very high probability of a disaster hitting you, one whose consequences could range from moderate to major, just how would your planning be different?

So what do meteors, auto accidents, floods, and hurricanes have to do with the Y2K problem? We think you can see much of the answer for yourself. The difference is the previously mentioned disasters happen with relatively little or no advance warning. Very little can be done in specific preparation. But in every case, if people knew these disasters were going to happen they would take whatever actions they could to prepare and minimize or even prevent as much damage as possible.

We realize we are going into what some might consider overkill on this issue. The reason we are doing this is because there are still too many people out there who are not making any plans for any type of Y2K preparation. We believe this is a serious mistake on their part. And if we have to beat the preparedness issue into the ground even more to get those people to do at least some preparing, then a few more pages on the subject are justifiable.

Y2K is coming. There is absolutely NO statistical doubt about it. Outside of the Lord coming back before this date, the morning of January 1, 2000 arriving right on schedule is an absolutely verifiable fact. The unknown is not if it will have an impact on your life (and everyone else's lives), the only unknown is how much of an impact it will have.

We will put it a different way. It is our opinion, and the opinions of many that are far more qualified to say, that this event is probably the largest social and economic event that has ever been able to be so accurately predicted.

Period.

So just what are you going to do about it? To do nothing should not even be considered as an option. Doing nothing will be an extremely unwise choice.

Why you should prepare-no matter what

There are three ways (Actually there are four, the first would be to do absolutely nothing.) to approach preparing for overall effects of Y2K. The first approach is to do what is the least difficult, least time–consuming, least expensive, and least disruptive to your lifestyle, yet offers you the most preparedness payback for whatever efforts you apply to it. On the other end of the spectrum is preparing for the "total self-sufficiency" approach. It is the most work, most time–consuming, most disruptive, and most expensive-both short-term and probably long-term. And of course there is the middle approach that combines some of both. It is that middle approach that is our top recommendation for how you should prepare. We will go into more detail about it after the section on covering the basics with the "most for the least" approach.

Even if you are STILL not sure about this Y2K thing and if it will have much impact on your life, there are still a number of practical preparedness steps that you can take with minimal expense and minimal trouble.

As you read through this first section on doing the minimal preparation for the least amount of money and trouble, keep this in mind. It will have the least disruption to your normal lifestyle and wallet to do the things we suggest in this particular section. Or looking at it another way, if little or nothing comes out of this whole Y2K problem, having done the things in this section will not have caused any major trouble, expense, or detriment to your life. Since doing these things won't cause you a significant amount of trouble, it is better to do them and be at least somewhat safe, then to do nothing at all. There just isn't any good reason NOT to do them.

But first, some really good reasons for preparedness even if this whole Y2K thing didn't even exist.

Preparedness is a good thing.

(Author's note: We are throwing this section in for those last few holdouts who are *still* not convinced they should include any kind of preparedness in their lifestyle.)

It was Duane Moll of The Urban Homemaker who said to us, "It is better to be prepared and red faced, then to be wrong and unprepared." In the case of Y2K, we agree with Mr. Moll's assessment 100%. It is far better do a lot of preparing, and have nothing happen then it would be to do little or nothing to prepare and be caught in potentially very severe consequences due to Y2K related problems. This is a clear case of weighing the risk vs. the probability as described previously. And in this case there is BOTH high risk AND high probability. Based on all the evidence, the *greatest* risk would be NOT to prepare for at least *some* disruption to your lifestyle.

Let us step away from the whole Y2K issue for just a moment. Without wanting to sound like paranoid, alarmist, doomsday type people, let us look at some other potential reasons to want to be prepared for disruptions of our traditional lifestyles.

Acts of God

There are plenty of natural disasters, or "Acts of God" as insurance companies and other businesses like to refer to it. (*Funny, isn't it, how so many non-Christians, and those who don't believe in God, will acknowledge God's existence yet refuse to have a personal relationship with Him, or allow Him to have any affect in their lives. They realize He is able to control the entire Universe from the stars all the way down to holding the atoms together, yet they won't let Him have any control in their life.*) Every month, we hear or read about these natural disasters bring major disruption to peoples lives somewhere. Just look at all the disasters that could potentially have an effect on your life.

Early in 1998, before our first Y2K conversation, Tammy (the wife) told Dan (the husband) she wanted to buy, of all things, a generator. Now we realize most of you can imagine how a *typical* husband and wife conversation about buying a generator might go: The *husband* says to the *wife* "Dear, I would really like to get a generator." And. of course the wife asks the classic question "Why, what do you need a generator for?" The husband gives what he thinks are a reasonable list of ideas, most of which the wife thinks are either impractical or totally unnecessary. Well, in our case the conversation went something like..."Dan, I have seen too many news reports of the ice storms in Maine and Canada. The idea of being without electricity for weeks on end in the dead of winter is not an option for our suburban home. With Minnesota winter temps hovering at 0 degrees down to -30 degrees in January, no electricity would quickly become a matter of life and death. I want to be prepared."

Since money was extremely tight, and Dan did not see a generator as being a high priority whatsoever, we did not buy a generator.

Guess what happened not once, but twice within months of that conversation? We were hit by two incredible storms, both, which knocked out the electricity for the first time ever in this house! The first time was for only about four hours, and our major inconvenience was being forced to go out to eat supper somewhere that did have electricity. (Mom says, "Yeah! Pizza!) But the second time we lost power was different. A major thunderstorm swept through Minnesota, leaving nearly half a million people without power. We were among the "lucky" ones. We were only out of power for about 16 hours. There were over one hundred thousand people without power for over 24 hours, and tens of thousands of people who were without power five days or more.

Guess what we are highly motivated to buy now? And we would probably want to buy it even without the Y2K issue looming ahead. Between the two (losing power and Y2K), we now have this strange new desire to have a generator in our garage!

Look around us in the last few years. There are many situations where having some advance "disaster preparation" of any kind was very wise.

- Ice storms in Maine No power for weeks

- Florida fires thousands forced to flee with virtually no advance notice.

- Floods in the plains states destroy entire towns, isolate thousands

- Tornadoes destroy entire towns every spring

- Past hurricanes can virtually wipe out entire communities and as in the case of Mitch in 1998, virtually an entire country!

- Potential earthquakes and other disasters. Earthquakes are possible in almost every major city in the USA.

A person does not have to believe that civilization as we know it is coming to an end, to be motivated to have some amount of disaster preparedness. Just picking up a newspaper or watching the evening news is enough to show us there have always been times and places where being prepared has had a major difference in the lives of those who were prepared for the unexpected. Those who are completely dependent on the electric company for their survival stand to lose the most.

Look at what happens in the case of a hurricane, as we discussed a few pages ago. (You may feel we are rambling about disasters too much here, but the points are actually extremely relevant to Y2K. Especially like this next point about people scrambling at the last minute to prepare for a hurricane. Take this next example and magnify it worldwide. You will get the picture.)

Acts of Man

As previously mentioned, unlike many natural disasters, like tornadoes that give essentially no warning, people have some advance warning of a hurricane and can do some preparing. But examine the preparations a little closer. If you look at the whole process, you realize the *lack* of preparedness by most people results in a frantic last–minute buying frenzy at the store. Store shelves get wiped out and become empty of virtually all essentials. And, as a result, many are forced to go without. The stores just don't stock enough merchandise on an every day basis to supply everyone with all the essentials at the same time. It goes without saying (but we are going to say it anyway) that those who are prepared in advance don't have to face going without. And as an added bonus, those pre-prepared don't have to pay the exorbitant mark-ups of opportunistic people selling things at sky-high pricing, seeking to make the most financial gain out of panic buying. The more you think about it, the more you will realize what we are talking about.

And something bigger than a hurricane is coming our way. And we know exactly when it is going to hit. Even if there are absolutely no product shortages due to Y2K –related problems, there may be product shortages just due to Y2K hysteria–induced panic and pre–event hoarding. Do you want to risk standing in line with the masses on December 1999, trying to get your hands on whatever might be left? And don't forget, if you do find what you need, it will be "limit 6 per customer" or 2, or even 1! And the prices! Who could have imagined? Get the picture? Would it be better to start stocking up now or then?

And if "Acts of God" aren't enough to motivate you to have as least some preparedness for your family, how about the ever disruptive "Acts of Man"!

Hardly a family in this country managed to remain completely unaffected by the UPS strike of 1997. For some, like our family, it had a crippling effect. In our business, The Home Computer Market, it was impossible to get software from some of our suppliers. So many of the homeschooling families who ordered from us, had to either wait until we found software from other suppliers or wait until the strike was over. Fortunately, we weren't dealing in perishables or things essential to life, like medicine.

Look at some of the other recent strikes:

- The GM strike had an enormous ripple effect on not only its now idle employees, but also all its suppliers. No longer could they ship parts to make into cars. Their businesses suffered as well. The GM Corporation lost millions of dollars in revenue. Sale prices of existing new and used cars went up due to unsatisfied consumer demand for cars.

- The Northwest Airlines Strike just recently also affected thousands of people. Not just the employees of NWA but those who use the airline to conduct business and essential duties. In our business alone, all our Priority Mail shipments slowed down considerably because all airmail in the Minneapolis area flies out on NWA!

- The US West Strike. Anyone needing a phone installed were told they would be waiting 4-6 weeks for an install! People moving into new homes would be without phone services for weeks.

These strikes not only affected the companies and employees of the companies, but the ripple affect in one way or another affected millions. Now, we realize that these are not going to wipe your house out or possibly disrupt electricity but they can affect those who are dependent on these systems. Losing a job, a sudden downturn in sales and profits, or even missing a shipment of essential items can make the unprepared panic and lose hope. But what if you had prepared by having extra supplies on hand? A year's supply of food and toiletries? How devastating would these man-made disasters be now? we think most of us would be better testimonies of God's provision if we followed Joseph's example and used the years of plenty to provide for the years of lean. Not only do we provide for our own household or business, we have plenty to share with others in need.

You can never predict other possible "Acts of Man." Just consider things that have really happened, that no one could have predicted.

- Terrorism
- The Oklahoma Bombing
- Wars
- Strikes of all kinds
- Lay-offs
- Accidents of every assortment
- Riots

Even more common are "personal disasters" such as death in the family, divorce, losses due to crime or fire, unexpected unemployment, loss of home, disability or illness, crime, and so on. Think about how any of these tragedies could be more easily handled if you had only prepared. Store up during the years of plenty to provide for the years of lean.

(Now, we realize that for many of the readers of this Handbook, we are "preaching to the choir." Many realize Y2K is coming and the need to prepare. We understand this. We have tried to cover the issues of the causes of Y2K problems and warnings to prepare in sufficient detail to help those of you who are still NOT so convinced, or at least until reading this Handbook, to become aware and the prepare. So now, without further delay or interruption, back to our previously scheduled preparedness advice!)

CHAPTER 9

THREE LEVELS
OF PREPAREDNESS

But first, a different view

Before we go into the specifics of what preparations to make for Y2K, there is a related issue we feel strongly lead to address.

This is a major concern that we want to share with you about as you go about investigating this whole Y2K issue. Many of those "beating the drum" regarding this issue have a significant vested interest issue. You need to be aware that many of those people leading the rally cry of "Prepare! Prepare!" have a bias. As part of their advice, they are trying to sell you things they recommend as part of the suggested preparation.

They make a profit if you follow their advice, whether or not the advice is good or bad. They make a profit if you follow their advice, whether or not you actually need what you buy from them. They make a profit if you follow their advice, even if you never use what buy from them. Therefore we question just how objective their advice can truly be, since they have a financial interest in what you do to prepare for Y2K.

The whole Y2K preparedness thing has become a big business, all in itself.

Now, it could be argued that we are guilty of the same thing. But not in the same manner. Yes, we are trying to sell a book-this one which you are reading. But this is all we are trying to sell. (And, of course, we think you should buy lots of additional copies to share with your friends and neighbors!!!)

"Little House on the Prairie"
or
"Nightmare in the Middle of Nowhere."

Something, which troubles us even more is that these people rarely ever give you any kind of truthful disclaimer or warning of the possible negative consequences of following their advice or the total real costs, involved.

We believe giving the type of advice we are seeing given, is completely irresponsible if it is not followed up with potential negative consequences or side effects.

To us, for someone to suggest something as drastic as "sell your house and move out to a very isolated area" is incredibly impractical and irresponsible if that advice does not address several critically important issues. Even if you are just renting, very few people could actually do it. For starters, very few people could actually pull this off successfully. Most who recommend everyone move to the country and become totally self sufficient, totally ignore the fact that most people lack the skills and abilities required to succeed in even the most basic tasks requires to achieve the level of self sufficiency being recommended. Secondly, it takes a great deal of time and planning, years perhaps, to become completely independent of the "world's system."

The advice to 'move away from civilization' usually never addresses all the various issues related to such actions. Issues such as the full costs involved, lost job and lost income, how to produce a new income or find a job in the middle of nowhere. If they use the argument that you won't need a job because you will be living off the land and your expenses will be very small, we suggest you re-read the previous paragraph. As idyllic as it may sound, the "Little House on the Prairie" fantasy is completely unrealistic for the vast majority of people. For most people "Little House on the Prairie" would turn into "Nightmare in the Middle of Nowhere!"

A Christian's responsibility

And none of this addresses the issue of a Christian's responsibility to respond *as a Christian*. What would be the consequences of your actions to your neighbors? How would your actions affect your witness to each and every one who knows you? Not just your lack of ability to witness. How do you be salt and light to the world, if you flee the world? Also, what is the unspoken message conveyed to them by your actions? (Remember: "Don't be a wacko!"). Keep in mind, as Christians, our actions speak far louder to those who know us, than any words we can ever tell them. We can ruin our testimony when others perceive us as "wackos." Let me quote an apparently non-Christian who has an opinion of all the hysteria surrounding most Christian Y2K publications: "It's been obvious to me for quite some time that the rising noise levels from extremist viewpoints would soon get in the way of moving forward with needed Y2K efforts. Although it may serve the needs of some to decree that Y2K is a sign from God to repent and prepare for the Rapture, I am not of that extremist camp."[121] He is right. Extremist survivalist recommendations cloud out the best plan of attack: the best witnessing opportunity of the century!

Even the cover of Time magazine shows the world's warped view of Christian believers. The January 18, 1999 issue shows a Christ–like figure, with his back to the viewer, holding a cross in his right hand, and a sign on his back saying, "The end of the world!?!" The articles are filled with "Christians" and "extremists" who are not only expecting the worst but hoping for it. Stockpiling goods, moving out to a remote area, and getting rid of modern appliances. We would much rather see news reports of the Church preparing their members to serve in times of crisis, preparing their facilities to hold cold and hungry refugees, and preaching the gospel. These types of Armageddon prophecies do absolutely nothing to promote the true message of the cross. When Armageddon truly does come, no one will listen.

We believe, if you move to the country, the most important question is this: Where would you attend church? We believe the issue of what church a person attends is actually of greater importance than their response to the whole Y2K issue. And that is why we feel extremely compelled to do significant coverage of this issue, and we dedicated one whole chapter to it.

For our family, this issue and this issue alone is the single main reason why we live where we live. So we can continue to go to our present church.

As you use the rest of this Handbook , we would like to give you these suggestions to get the most out of it. Not everything will apply to every family. Everyone will bring their own situation to these things. Some will apply, some will not apply. Many of these suggestions are not things that we are saying you should do. They are things you need to address and decide for yourself. Of the areas you are looking at with interest, you will probably want to do more research.

As you read the sections on preparedness, remember that if you plan on preparing at the "greater" preparedness level strategy, you will want to make sure that you also do the things at the "minimum" suggested preparedness level in addition to the suggestions at the "greater" preparedness. And, of course, if you plan for the worst, make sure you are covering the appropriate items in the previous two preparedness sections.

Minimum Suggested Preparedness: "The Most for the Least"

Even if you are not sure about this Y2K thing and if it will have much impact on your life, there are still a number of practical preparedness steps that you can take with minimal expense beyond what you would ordinarily spend anyway and with minimal hassle.

As you read though this first section on doing the minimal preparation for the least amount of money and trouble, keep this thought in mind. If you do all these things we suggest in this particular section, it will have the least disruption to your life. In fact, most of these suggestions are beneficial regardless of Y2K. That is why we put them here. If little or nothing comes out of this whole Y2K problem, having done the things in this section will not have caused any major trouble, expense, or detriment to your life. Putting it another way, since doing these things won't cause you too much trouble, it is better to do them and be at least somewhat safe, then to do nothing at all.

There just isn't any good reason NOT to do them.

Larry Burkett, of Christian Financial Concepts, has these basic recommendations for basic Y2K preparedness. Have on hand at least:

- 4-6 weeks worth of food

- Bottled water

- Alternative heating source. He recommends a stand alone oil heater, one that is designed not to be vented.

- Oil lamps

Best Case Scenario Preparations

These are our steps for "minimal preparations" and are in anticipation of the best case scenario. This assumes these factors:

- You have electricity or if you lose power, it is only for 48 hours or less.

- Food shortages are minimal. We feel the chance of food shortages is much higher than the chance of losing our power for an extended period.

- You do not lose your job or primary means of income.

- Your water supply is not affected.

Step One:

Begin buying to establish what James Stevens calls a "safety net." Safety net purchasing involves beginning immediate buying of "extra cans, jars, or packages of foodstuffs, medicines, and household products you routinely use." He suggests two methods to this "planned co-buying:"

- Dupli-buying: Buy 2 of any food item you usually buy. This does not include perishable or low priority items.

- Multi-buying: Buying large quantities of anything you use that is on sale. we will expand on this in a bit.

You can start this type of buying immediately with minimal

disruption to your budget.

Step Two:

- Stock up on other essential items such as water (2-Week supply), toothpaste, diapers, etc. Determine quantities based on your chosen level of preparedness. Specifics are listed beginning on page 178.

- If possible, have on hand a sufficient sum of cash for at least one month of expenses.[122]

Now, let's look are some very specific examples of this type of preparedness. How about, and we hate to even bring up such a delicate subject, *toilet paper* (or "Bathroom Tissue" as the manufacturers and stores, but no one else in the real world, likes to call it). How would it affect you if you had none and had a hard time getting any? How much would you be willing to pay for it, if it were in extremely short supply?

Now let's look at the other side of the *toilet paper* issue. How about this scenario? What would be the downside if you *did* stock up a 1–3 month supply of toilet paper and when January or February 2000 comes around and all the stores still have toilet paper? Would you have to throw it all out? Would it suddenly become useless? No? Oh, you need to have some around anyway. So you would just temporarily have an excess supply. And you may have some temporary minor inconvenience of the storage of excess supply along with other overstocked items.

But look at the bright side of the toilet paper issue. First of all, you won't have to waste time running out to the store to get toilet paper for a few months. Second, if you strategically and methodically did all your stocking up, you probably will have bought it all when it was on sale for the cheapest price you saw it at in the last three months. So it will have been a frugal purchase and a good investment, since you would have had to buy it all sooner or later.

The Alpha Strategy

The first time we read about this type of saving though stocking up on bargains, it was in a book called *The Alpha Strategy*. The book was printed sometime in the late 1970's or early 1980's, a time of high inflation and many new investment strategies. We read it in 1985. At the time, we were doing research on economics, gold, silver, investment strategies, speculation, the history of investing, inflation, currency, and the economy. Of all the research we did, *The Alpha Strategy* made the most sense and is the one thing that has had the greatest and most beneficial impact on our finances. Basically, *The Alpha Strategy* entails purchasing large quantities of items you normally use at bargain prices.

In fact, as we look back on it, we can hardly recall a time when the approach of strategic stocking up on bargains turned out to be a bad thing. (Although there was that one time when the huge bag of ripe bananas turned out to be not as good of a deal.) Aggressively stocking up, in quantity, on highly perishable food items is not such a good idea unless you can or somehow preserve your fresh produce as soon as you get it.

The other factor to consider is the savings of buying in bulk. While buying in bulk does not always guarantee the cheapest price, it usually will offer savings over similar items bought non-bulk. Buying in bulk entails buying the largest quantity possible given your family's diet and your storage capacity. Bulk means buying 25 pounds of oatmeal at 25 cents a pound rather than one pound in the fancy container for $1 a pound. Of course, this type of volume buying is always best if you have 7 or so hungry kids at home.

Tips on where and how to buy in bulk:

- Check around at church to see if anyone is involved in a local food co-op. These are groups of people who buy large quantities of food at near wholesale prices and divide among the group, splitting the cost. Many co-ops specialize in organic or natural food.

- Start your own co-op.

- Go to your local grocer and ask for quantity discounts.

- Check out your local Sam's Club or other similar wholesale warehouse club type store. Many bargains can be found at Sam's but be aware, some items are not any cheaper than the local warehouse grocery store. Price each item carefully.

- Visit a local "farmers' market" usually held weekly on Saturday mornings at a specified location. One step better: get to know and do business with one particular farmer. When tough times come, regular customers will get preferential treatment.

- Check restaurant supply stores to find large cans of veggies, beans, etc. and other bulk items.

- Find or establish exchange or barter organizations to trade your surplus items or labor for items you need.

The real exception to this is when items are on sale as "loss-leader." For those of you unfamiliar with the term "loss leader", it refers to those items stores advertise at drastic discounts in their sales ads. Often, these are those things the store lists as "Limit 2 per customer" in the fine print. They know that even though they will lose money selling these items below cost, they will make a healthy profit on everything else in your shopping cart. So overall, their small individual product losses turn into higher overall profits. Whenever we see these sorts of things on sale, if we need it, we will buy up to a 6-month supply provided they aren't "Limit 2 per customer."

It was *The Men's Manual Volume II,* published by *The Institute in Basic Life Principles* which taught us to view both spending and the savings from discounts from some very different perspectives.

When people see the price of something listed at $9.99, they usually fail to factor in various components that are very relevant to the real cost of the purchase. A $9.99 purchase actually costs about $10.50 to $10.80 with the sales tax that is also due on the purchase. But before you can spend the $10.50 to $10.80, you have to earn it. For many people, having this much money to spend requires earning $13.00 to $15.00 BEFORE taxes.

Whereas most people can begin to see the obvious advantage of saving 20-30% (or more), there is another angle that we never thought of until we read about it in *The Men's Manual Volume II*: the other side of the savings equation.[123] If you can save 20% on that same $9.99 item, that is like investing your money at a better than a

20% return on your investment-guaranteed! Their philosophy is that you need to look at savings in terms of percentages, not dollars and cents.

Looking at this in terms of bulk or multiple purchases, it is like buying 4 and getting 1 free or getting 25% free. If the discount is larger, the saving factor goes up significantly. If you can get something for 35% less, it is the same as getting 50% more for the same amount of money! Example: If something normally costs $1 for a certain amount, if you can find it for 35% less somewhere else, you can get over 50% stuff more for the same amount of money. Here is the math. A 1 pound box of Acme FoodStuff costs $1 at your local grocery store. But you find a source to buy the same thing in bulk for 65¢ a pound somewhere else. The same $1 can now buys you 1.53 pounds of Acme FoodStuff. $1 divided by 65¢ per pound equals 1.53 pounds. Over 50% more Acme Food Stuff for the same money!

Take this example one step further, and look at the other side of the tax advantage just mentioned. When you earn money, you have to pay taxes on the money you earn, before you even get the money to spend. But you don't have to pay any taxes on the money you are saving when you buy something at a discount! If you invest money and make 10% interest, you have to pay taxes on the money you earn. (Unless, of course, you invest in tax-free investments. But here, the investment return is almost always at a lower rate than similar types and risks of taxable investments.) So if you save 35% on something, and you are at a 25% tax rate, you are getting true rate of return on your money that would be similar to a rate of 72% after taxes! (This 25% tax rate estimate is a conservatively low rate when you factor in Social Security, Medicare, Medicaid, state taxes, sales tax, and every other tax we pay) you The math works like this: Take that same $65.00 and invest it. Now, you need to earn enough money on that $65.00 to be able to buy the same $100.00 worth of goods (remember, this is the regular price, not the discounted savings price.) Plus you also have to earn enough to pay your taxes on the profits. $65.00 at 72% equals $46.80. Taxes on 46.80 come out to $11.70, leaving $35.10 profit. Now take the $35.10 profit, plus the original $65.00 and you have $100.10 to go out and buy the same goods you can buy for $65.00 at a 35% discount. To say that a 35% savings is the equivalent of a 72% after-tax return on your money sounds impossible but that is how it all comes out! Put in you own tax rate in the formula. Don't forget to factor in all those little additional taxes you pay.

This example shows just how great the financial power of the Alpha Strategy can be. As a general rule, take the percentage savings, double it, and that is sort of what your after-tax equivalent return on your money will be.

When you look at it this way, stocking up appears to be not only a wise Y2K strategy, but also a fantastic investment of your money. How to plan your in home food storage will be discussed in a later chapter.

Waiting until the last minute

Someone had brought up the issue of just waiting until November or December 1999, heading out to the local discount stores, and doing a massive stocking up then. And if January 2000 comes with little or no impact, just return most of it.

Two thoughts on this strategy:

First, the risk factor.

Do you really want to gamble your future needs for essential goods on the hopes that everyone else is not doing the same thing? You can be sure of this, even if little or nothing "stops working" in January 2000, there will most likely be at least some level of a panic buying frenzy in the fall and winter of 1999. As 1/1/2000 gets closer, the general public will most likely not want to be caught without their own essentials. Unless manufacturers and supplies work way overtime and put production and workforces at full capacity, we predict massive spot shortages of many items due to hoarding. And any essential goods left on the shelves come November or December 1999 will have a high probability of being available only in limited quantities per customer. Do you really want to take the risk of putting off preparing and storing until the last minute? We certainly hope not.

Second thought on this issue.

Ethics.

Let us look at the ethics of buying large quantities of merchandise, with a high intent to return most of it if everything seems close to normal in January and February of 2000. From our values, we see this as morally and ethically wrong. And we think that same viewpoint would be held by many reading this Handbook .

Yet we will take this issue one step further. The 8th Commandment. What does returning things you bought legitimately have to do with "Thou shalt not steal?" A lot, if you look at your intent. Look at what happens after you walk in the door to return something. How many people have to handle that return merchandise before it gets back on the shelf? How about the accounting that needs to be done to properly keep track of the money flow back and forth? As someone who has owned several small businesses, we can tell you it can be a major headache, hassle, and expense. If you are just buying a bunch of merchandise on pure speculation, you are gambling with the stores money and time. And many would say that ethically, you are stealing that time and money from the store. Whether or not it is your "intent" to be stealing does not change the issue or lessen the cost or trouble to the store if you end up returning a large amount of merchandise to the store.

We just don't think the strategy of waiting until the last minute is a wise one, a right one, or one worth the risk. End of issue.

Financial strategies.

Cash.

Another area of preparedness that fits into the definitions of minimal, easiest, least costly preparing, along with "The MOST bang for the least buck" is cash. As in green back, and also known as Federal Reserve Notes. Think of that so-called "real" money stuff that you go to get out of the ATM machine. We are NOT referring to what is more loosely known as money or dollars, which is be found in the form of checks, credit card transactions, money orders, or electronic funds like what is "sitting" in your bank account. We need to be specific here, so there is no confusion.

As we look at the issue of cash and Y2K, there are two perspectives to look at. The first perspective is cash as it relates specifically to being ready for Y2K. That is what we will deal with here. The other perspective is from an investment angle, which we will deal with later in the Handbook.

If very many Y2K related glitches hit, banks will no doubt feel some effect. It is a very realistic possibility that some banks will be closed for days, weeks, or possibly even months. It is even possible that some banks could go out of business. (Don't worry, even if your bank goes out of business, the FDIC will insure for your money up to $100,000. It may take weeks to get your money out, but you will get it!) ATM machines may not work. Bank accounts could be frozen. Credit cards may not work. You may lose your job.

You may have thousands of dollars in the bank, and that is where it could end up staying, in the bank, where you can't get at it. Yes, your bank is FDIC insured, so you will not lose the money in the long run. You just may not be able to get any of it out for weeks or (worst case) even months. A bigger bank would have a lesser chance of closing or failing. Another tip: spread your cash assets among several banks. With that said, let us challenge you with this question, "How long could you survive on just the cash you have on hand alone?" No checks, ATM cards, credit cards, or such. Just cash.

Having this as the perspective, we believe EVERYONE needs to have, as a bare minimum, at least a 30-day supply of cash on hand. This is what we mean by a 30-day supply. Have as much cash on hand as you would normally spend in your total monthly expenditures for everything in a given month. This includes house payments or rent, car payments, food, electricity, telephone, gas, repairs, heat and whatever else. Do not forget to include those things that might be automatically withdrawn from your paycheck, like life or medical insurance. Also consider having some extra cash to give to those who need extra help.

We actually suggest you have a 3-month supply of cash on hand, but even a 30-day supply will be hard for many that find it hard enough to even make it to the end of the month with any money at all. Keep in mind, however, there are federal requirements on how much cash you can withdraw or deposit. If you withdraw $10,000 or more, you must fill out a special form. If you withdraw smaller amounts over a period of time to avoid this requirement you could be in trouble. Talk to your bank advisor about these rules.

Now, let's look at the potential downside of following our advice. You have all this cash on hand and when January 2000 hits, nothing happens. Banks stay open, businesses are fine, and things are just like the month and year before. No problems. So you have all this cash and no crisis to spend it on! Does this mean that now, since there is no crisis, all this cash is no good now? Do you have to throw it all away now? Obviously not. You can still spend it or put it back in the bank. Worst case scenario is you lose a few months of interest.

In anticipation of this new demand for cash, the Federal Reserve is already, *for the first time ever*, planning to boost currency reserves by $50 billion to $150 Billion, in addition to the $460 billion already in circulation. The Fed is estimating that each of the 70 million households will, on the average, withdraw $450 to pay for necessities, such as food and gas. Other contingency plans include adding workers at the Fed banks, printing larger denomination bills such as $50 or $20 instead of $5 or $10 bills, and allow older bills to remain in circulation.[124]

Debt

We would highly recommend you reduce or eliminate as much debt as possible. Because there is so much information of the advantages of this, even without the Y2K issue, we won't go into many details on the obvious advantages of this. Specifically in terms of Y2K preparedness we will say this. If the economy does turn down into a recession, money will be tight and people will be losing their jobs. This is the worst situation to be in and have debt hanging over your head.

Since we are on the subject of debt, there is something we want to clarify. Someone who did not understand the Y2K issue very well brought up the following idea: charge all kinds of stuff in December 1999, because the credit card companies computers will all be crashing and they won't be able to track it. Wrong.

Aside from the obvious (go back a few paragraphs, regarding the 8th Commandment) we think it is a financial mistake. If we had to bet (which we do not, because it is wrong) on whether or not the credit card companies will be Y2K compliant or not, we would bet on them being compliant. Many already are. Don't you gamble, and don't you steal. Even preparing for Y2K doesn't justify compromising your convictions. That one should be clear enough.

Lastly, we do not recommend going into debt to fund your preparedness plans. If the Lord knows you need three months of food, He will be able to provide the money. Otherwise, others will have the opportunity to minister to you by sharing out of their abundance. Rest in God's perfect timing and provision.

Other Basic Y2
Preparedness Tips

Check out your own computer

While this Handbook was not designed to help you overhaul your PC, we did want to give you a few tips on getting your PC Y2K–ready.

- Macintosh computers are already Y2K compliant but their individual applications may not be. Check with individual software vendors for their Y2K readiness.

- Neither Windows 95 nor Windows 98 are fully Y2K compliant. Check out Microsoft's web site for software upgrades and information.

- The older your machine, the more likely you will have problems. "PCs with 286, 386, 486 and older Pentium microprocessors have battery–powered CMOS modules with built–in clocks and calendars— most storing only the last two digits of the year."[125] Many of these modules will have to be replaced. Another potential problem is the BIOS. All should be checked even though most installed in Pentium–based systems manufactured in1995 and after are compliant.

- Embedded components such as BIOS chips, motherboards, applications and operating systems are all capable of failing and must be checked, tested and/or replaced.

- Try contacting the manufacturer of your PC but the best bet is to test your PC yourself.

- Computing Today magazine recommends this tip: "One solution is to get your hands on Norton 2000, a $49.95

Three Levels of Preparedness

program from Symantec Corp. (www.symantec.com). The software is a one-shot Y2K checker to fix a bad PC BIOS, flag bad applications and data files. The software allows you to test the BIOS using a "boot floppy" thus avoiding any potential damage to your system. A "smart" reading of dates in data files lets Norton 2000's checker rank potential errors in order of severity, and the program will feature an Internet update to include some form of automatic date correction for data files, thus saving hours of manual labor."[126]

• Another web site with Y2K diagnostic tools is www.msnbc.com/news/TECHY2KTOOLKIT_Front.as

• Check out your church's computers.

Get hard copies of all important documents

Some experts recommend getting hard copies of every document pertaining to your financial, dental, and medical records. Examples include:

• Social Security records for every worker in the family

• Credit reports from the big three credit bureaus, TRW, Equifax, and Trans–Union.

• Bank Statements

• Proof of payment on bills

• ALL tax records for the last 7 years

• Birth certificates for each member of the family

• Marriage license or certificate

• Deeds, titles, or other proofs of ownership

• Mortgages and other loan documents

• Credit card statements

• Baptismal, confirmation or other religious documents

Larry Burkett recommends a few other preparedness actions"

• "Write [or call] your banker, asking for a written assur-

151

ance that your bank will have all its internal computer systems reprogrammed and tested by the end of this year, leaving all of 1999 for external testing. If you can't get that assurance, switch to a bank that can give you that assurance.

• Keep printed copies of all financial transactions- bank, insurance, money market fund, mutual funds, stocks, taxes, etc. Also get a copy of your Earnings and Benefits Statement from Social Security. You may need these hard copies to verify your financial claims if computer-based information becomes corrupted.

• Write [or call] your utility companies (power, water, natural gas) and request written responses about their Year 2000 compliance strategies.[127]

The Second Level of Preparedness
(This is the level we recommend.)

This "second level of preparedness" is in anticipation of the "most likely" case scenario. This assumes these factors:

1. You may have electricity and if you lose power, it will be only for 2-7 days.

2. Shortages of most items are likely but last only for a week or two.

3. You may lose your job or primary means of income

4. Your water supply may be affected.

5. You may want a 1–3 month supply of your basic needs

Your Survival Priorities

At this level of preparedness you will need to firmly establish your survival priorities. They are shelter, water, fire and food in that order. This list is from a company that offers preparedness workshops.

"1. The most important survival priority is shelter. You need to be protected from the elements - sun, wind, cold, rain and even insects. We are particularly susceptible to the effects of cold, wind, and rain. You will die in a matter of minutes to hours depending on the temperature. Once hypothermia sets in the vital organs begin to shut down. The brain is first.

2. Water is your second survival priority. Water makes up 75% of the human body and needs to be constantly replenished. Once you have fulfilled your needs for shelter, all effort is concentrated on finding, gathering and treating water. Having some stored water on hand in case of an emergency situation makes good sense.

3. Fire is your next priority. It provides warmth and light, the ability to cook your food and a way to treat your water. But fire also warms the human heart and goes a long way to making you feel more secure.

4. Food is the last of your four basic needs. You won't starve to death if you don't get three square meals. Food encompasses everything from plants, animals, traps, weapons, learning to hunt and gather, cooking and storage. In preparing for an emergency situation we do advocate storing food. It is only prudent."[128]

How to Deal With Various Y2K Induced Problems

Electricity is the big one. If the lights stay on, MOST of the doomsday predictions WILL NOT happen. Re-read the section on the utilities on page 28. Read our reasons why we believe we will not lose power. However, if the electricity goes out for longer periods of time and if these outages occur over a widespread area, i.e. whole cities and states, one can expect serious problems would result. Darkness tends to attract evil and without lights, our society would be in mild to major panic. Several essential services would be directly affected by the lack of electricity: HEAT, WATER, AND LAW ENFORCEMENT.

HEAT

Your first concern is providing heat to your home or shelter. If you live in an extreme northerly climate, i.e. Minnesota, this may be even a greater concern than water initially. With January 1, 2000 temperatures possibly hovering near 0 F or even lower, staying warm quickly becomes a life and death situation. With the electricity off, even with natural gas as your source of heat, your furnace would not function. To keep your family warm and the pipes not freezing, the interior of your home should be kept at least 55 degrees. Of course if it is near or above freezing outside, there is little chance of your pipes freezing if your home's interior is lower than 55 degrees.

Your strategy for keeping warm depends on how long you feel the power will be off and how cold it is outside. While many are predicting the long term if not permanent demise of the power grid, we personally believe that any outages will be infrequent and of short duration. That does NOT mean that they can't happen to you. Everyone should always be prepared to weather the cold in case of unforeseen emergencies. Here are some short–term tips to keeping warm.

Make sure you always have a battery operated radio so you are able to listen to radio announcements. If there is widespread power outage, your community may already have contingency plans in place to open shelters for those in the community without power. These shelters would most likely be in a high school or other public building. Check out community preparedness tips at the end of Chapter 8. Help your community make these plans if they have not already done so.

Dress in layers. As a matter of preparedness, you should have bought all family members extra socks and underwear, a high quality, sub zero sleeping bag, long underwear, a high quality winter outerwear. "The layering system is recommended for cold weather conditions. Three items to include in your personal wardrobe are "expedition weight" Capilene underwear, fleece outerwear, and a protective outer shell of waterproof material."[129]

Insulate the layer between your skin and clothes if necessary. "If you are caught short in any emergency situation, a tip for keeping warm: insulate your clothing with the appropriate material as close as possible to the skin. In an urban setting you can use the pages from your phone book to stuff your clothes. In the wilderness, use forest debris, leaves and grasses. If your car breaks down, rip the stuffing out of your seats (remember it IS an emergency!) and stuff your clothes."[130]

Remember keeping warm requires a higher caloric expenditure by your body. Eat food with high carbohydrates and fats to give your body plenty of fuel to burn.

The elderly and small babies should receive priority when in comes to keeping warm. Their bodies do not regulate heat as well and they need to be checked often for hypothermia and frostbite.

Examine your fireplace. Can you place an insert within the fireplace to convert to a practical method of heating a portion of your home? Most fireplaces are not designed to heat, just to look pretty. In fact, a fireplace usually wastes energy due to the amount of heat drawn away from the house and spewed up the chimney. By placing the right insert in your fireplace and having a good supply of wood, you may prevent a serious problem if your heat goes out for a few days.

Some new ideas include vent–free gas heaters. New regulations have eased restrictions and sales of these economical heaters making them a good option for home heating. Jim Lord (A Y2K preparedness expert) points out, "My Tip of the Week is to consider vent-free gas heaters as, first, a great way to provide primary heat to your home. Secondarily, consider these heaters as a very inexpensive backup source of heat to your existing system in the event there are Y2K-induced power outages. Next, investigate the availability of natural gas in your area. If it is available, talk to the gas company and find out what the ambient pressure of the system is in the event electrical power is completely knocked out. You may find you can have all the gas you need. If natural gas is not available or if your system has insufficient ambient pressure, consider propane instead." He says these heaters can provide up to 30,000 BTU's and cost around $225-$400. They are safe and inefficient. Check them out. We are.

Examine your "shelter." If it is your home, plan now to increase your insulation, cover windows, add weather stripping and caulking to windows and doors, etc. Another way to keep warmer inside is to build a shelter within a shelter. Set up a tent in the living room with down sleeping bags and have everybody snuggle together. "Bear in mind that a home without heat is nothing more than a large tent. Pick the smallest room in the house that is safe, dry and the least exposed to the cold. Then make a shelter within a shelter. It is best to have a room without windows or with windows facing the sun. Pick a room that is not only convenient but also one with doors that can be opened for ventilation or closed to prevent a draft. Then go hunting! Gather up all the insulating materials you can find: blankets, pillows, mattresses, towels and clothing. Don't forget the drapes and carpeting. The padding under the carpet is excellent too.... And, NEVER bring flames or any combustibles of any kind into the shelter."[131]

Generators can provide temporary power in a feasible manner. There are natural gas, propane, diesel, and gasoline powered generators and generators that can use more than one type of fuel. There are generators that can run on either propane, natual gas, or gasoline. They are called tri-fuel generators. They will cost more, but they give you the most options and flexiblity for fuel choices.

Honda is one of the companies that makes a tri-fuel generator. There are also adapter kits available that allow you to retro-fit regular gasoline powered generators into tri-fuel generators. Here is one company that sell such kits:

- U.S. Carboration and Turbo Systems
  770-419-7698
  800-553-5608
  www.uscarb.qpg.com

When buying a generator, prepare to spend at least $500 for a minimum, $1,000.00 is recommended. If you do buy used, think of it like a used car. Try to find a mechanic who can check it out for you. A high quality, whole house generator will run $3,000-$4,000.

There are a variety of opinions as to which fuel is preferred. Diesel may be more readily available during a crisis and is easier to store for longer period of time. Gasoline is cheaper and you have the advantage of siphoning off gas from your vehicles if you are desperate enough. Propane is also cheap and easy to store but harder to come by during a crisis. Some communities without electricity may still have natural gas.

There are three features you will probably want to make sure the generator has before you buy.

- Name brand with good reputation.

- Automatic shutoff in case of low oil levels, to prevent motor burn-out.

- Cycle down feature. This allows the motor to slow down and use less fuel when it is not experiencing an actual draw or demand for electricity. This is ideal from keeping things plugged in like a refrigerator where the compressor turns on and off sporadically.

How will you use a generator? Practice before you need it.

Keep in mind however, by storing large amounts of such flammable substances as gasoline, propane, or diesel in or near your home may void your homeowner's insurance policy. An insurance agent friend of ours from our church urges caution before deciding to stock up on such fuels. Check with your own agent before deciding what to do. Some preparedness experts have cautioned against the use of generators unless you are very familiar with their operation due to potential for accidents or misuse.

Alternative heating methods will be discussed under long-term tips.

SUMMARY

In the worst case scenarios, the most pessimistic prophets (and we use that term loosely) see a near permanent shutdown of the electric power grid. We believe power shortages will be localized and of short duration and also keep in mind these fundamental facts:

1. The electric companies are just a bunch of primarily profit driven, greedy, capitalistic pigs. (We say this "tongue-in-cheek.") They have a high motivation to keep your lights on. We believe this "invisible hand" of capitalism will ensure continued and reliable electricity.

2. The government, as inept as it is, is also extremely motivated to keep the power grid up. There are powerful branches of the government (i.e. the military, the CIA, FEMA, etc.) who would make it their primary duty to get the power on and stay on.

3. Read the earlier chapter on utilities.

However, in the case of Dan and Tammy being wrong, you should have a contingency plan in place to provide heat to your family's dwelling in the possibility of a long term power outage (2 weeks+.) Here are some alternatives to consider:

Long term heating solutions

- Wood heat furnaces. For $4-5000, you can install a wood heat furnace that not only heats your house but your water as well. Some wood heat furnaces require electricity to function so you may need a generator as well.

- Generators. The major drawbacks to a generator are the need to procure and store sufficient quantities of fuel for using the generator on a continued, long-term basis.

- Solar panels. We have seen companies who install solar and back up power systems designed to run your whole house on solar power and backup generators for under $13,000.

- Windmills. In some areas of the country, wind power is a very serious alternative to any other energy source. In parts of Florida, Florida Power and Light will buy your excess electricity if you erect your own windmill. These payments are applied to your electric bill for times when you do need to access their power grid. (For example, days with no wind.)

- Move to a warmer climate. You may sweat more in the summer but at least you will not freeze to death in January. We wanted to escape the wintry prison of Minnesota but God told us to stay put, that he had called us to a certain body of believers in Minnesota, and that we were not to obey the flesh in this matter. Weigh any decision to move with God's purposes. Where will you fellowship? Will you lose contact and eventually your opportunity to witness to extended family members? Please read Chapter 16 on choosing a church.

WATER

No electricity would result in no water pumping, and no water supply would be a disaster. Water is essential for everyday life. Not many households can last long without any water coming to it. Here are some preparedness tips for ensuring your water supply:

• The average person needs a bare minimum of 1 gallon a day for basic personal needs, cooking and drinking. High temperatures, increased physical activity, and sickness would elevate the need for water. Nursing mothers and busy children usually need more as well. That does not include washing dishes, bathing, or washing clothes. For a short-term water supply emergency, some of these secondary uses can be eliminated.

• James Talmage Stevens in *Making the Best of Basics* suggest storing only a 2-week supply of water for your family. Because of the storage space needs and weight of greater supplies of water can be overwhelming, 2 weeks is a reasonable goal for water storage.

• Start storing water now. You can buy plastic canisters or drums commercially designed to store water (see preparedness companies) but there are cheaper options. Do you drink soft drinks? Buy it in 2 liter bottles. As you empty them, rinse, fill with water, and add 2 drops of bleach and store in a cool dark place. Do not use metal containers or lightweight, food grade plastic containers previously used for storing fruit drinks, milk, pickles, household cleaners or chemicals. Do not store your water near any paint products, gasoline, acids or other odor causing chemicals.

• Rotate your water supply to keep it fresh.

• Having bottled or stored water at any time is a key preparedness act. Just ask anyone who has been in a natural disaster such as an earthquake, tornado, or hurricane. There are few places on earth where none of those disasters are a possibility. Here in Minnesota, we do not have to worry about hurricanes or earthquakes but nearly every major city in the US has the potential of suffering a major earthquake and many cities are along coastal areas.

• Call your local water company and exhort them to have contingency plans in place to ensure continued water supplies. Mention the chance of electricity not being available and that emergency generators should be considered. Tell them that no water or undrinkable water would have dangerous outcomes.

• Examine your location. Is there a pond, lake, river or creek near-by? Plan your strategy now on how you would transport the water to your home and how you would make it safe for drinking. Portable water purifiers are available at most preparedness companies, sporting goods or camping stores. The cost can run from $200 to $500.

• Check into placing a well on your property. Make sure you have a hand pump available in case there is no electricity for your electric water pump.

• Don't forget the "built in" reserves of water already in your house. Your water heater holds 25-50 gallons, your pipes and system holds several gallons, and your toilet tanks hold a 5-7 gallons each. (Toilet tank water is not potable if you use commercial disinfectants or cleaners in your tanks.) If you feel really paranoid on December 31, 1999, fill your tub(s), water cooler, clothes washer, buckets, pots and pans with water and shut off the main water valve. This will protect your house's built–in store of water in two ways: one, it will prevent the city's water system from siphoning off your supply and two, it will prevent any contaminated water from flowing back into your internal system from the main water supply

• Tips for saving water: Use paper plates, cups, etc. (buy before Y2K). Make sure all the clothes are washed. Have extra socks and underwear on hand. Those are the hardest items to forego washing. Use water several times before discarding. An old Jewish proverb: "Never throw out dirty water until you have clean water."

• If there is no water, you cannot flush your toilets. Plan now on how you will handle human waste. Here are some options to consider:

Invest in a port–a–potty of some type. Found at most camping stores.

Lay a plastic bag inside the toilet bowl, use, twist–tie close and place in sealed container outside

Dig a deep hole in the back yard, four to six feet deep and a two feet square per seat for an outhouse and building a wooden toilet box. Toilet boxes should be approximately 16 inches tall. Make sure you dig it before the frost hits.

• Have on hand guidelines for treating questionable water sources and the chemicals needed. We recommend James Talmage Stevens' book *Making the Best of Basics.* Before treating any water, let any suspended particles settle to the bottom, or strain them through layers of paper towel or clean cloth.

Three ways to purify water:

> **Boiling:** Boiling is the safest method of purifying water. Bring water to a rolling boil for 10 minutes, keeping in mind that some water will evaporate. Let the water cool before drinking.

> **Chlorination:** Add two drops of bleach per quart of water (four drops if the water is cloudy), stir and let stand for 30 minutes. If the water does not taste and smell of chlorine at that point, add another dose and let stand another 15 minutes.

> **Purification tablet**: They release chlorine or iodine. They are inexpensive and available at most sporting goods stores and some drugstores. Usually one tablet is enough for one quart of water. Double the dose for cloudy water.

NO LAW ENFORCEMENT

This is the most "scary" to me of all the possible effects of no electricity. Without electricity, your local police force would most likely not function well. If the power outage is short (less than 2 weeks), I would speculate that most police departments would muddle through. If the power outage is longer, the police will have more headaches than just no power. Civil unrest, increased crime as people resort to stealing and looting to provide for basic needs, people traveling to "markets' with large sums of cash or gold to barter for basic necessities, gangs terrorizing suburbanites, no way to communicate with other federal, state or local officials, etc. It would not be a pretty sight. At this point one would be considering one's escape route or "plan B" options. Fleeing such mayhem would be a good idea for most families.

As a part of preparedness, you should have carefully planned such escape contingencies. Personally, our family is planning to hide out with Grandma and Grandpa in the northern woods of Minnesota. With a large garden, septic system, and well in place, they are already pretty self-sufficient. Keep in mind, these types of escape options are for the last resort only. A supply of gasoline, and a well-tuned vehicle may be required for a hassle-free get-away. We do feel that the most astute will see the warning signs and leave before it becomes too difficult. Re-read Chapter 6, "*Keep your finger in the wind.*"

GUNS

If civil unrest becomes more common, you may become the only law enforcement in your neighborhood if you are the only one with a gun. Before we get anywhere into the subject of guns, we want to make these two statements.

- We are not gun experts. In fact, at the time of this writing, we don't even own a gun. Never have. No, we are not pacifists. We do want the right to own a gun if we choose to. We just haven't chosen to. That is, until now. We are looking into our first gun purchase, and it may not be our last.

- We are making no recommendation either way as to whether you should get a gun or not. (How is that for wishy-washy?!?) We simply want to bring this subject up for each person to examine of him or herself. We do believe every family needs to examine the issue, seek good counsel, and pray about it.

With that said, let's discuss guns.

Many who are familiar with the Y2K issue are saying there is a very real chance there will be rioting and general civil unrest as a result of the problems created by Y2K failures. As a pre-caution, they are suggesting people buy a gun, start using it, and become familiar with it. Although we feel the probability of general civil rest is very small; we do want to discuss this issue.

We really do not want to get into the moral issue of owning a gun in this Handbook. As mentioned above, up until now, we chose not to own a gun. Many people we know own guns. In Dan's former occupation of rare coins and gold and silver, most of the coin dealers owned guns, and many of them carried them with them when they traveled. And with good reason. There was a very high robbery rate, and several times a year, Dan would hear of some coin dealer who was murdered in a robbery. While Dan was working at his first job in rare coins, that store was broken into twice. Thankfully, both times were in the middle of the night when no one was around. That same coin dealer also occasionally dealt in collectible guns, and usually had a gun with him whenever Dan and he traveled. And with all this, we never felt led to own a gun.

Until now.

To us, the potential risk to our family is high enough that the gamble is not worth not having a gun. Note, we say 'potential' risk. Today, there is little risk. But a person need only look back the riots in L.A. a few years ago to realize that civil unrest is a very real threat and possibility. And it would not take much for some areas to erupt again. Where we live, in a suburb of Minneapolis, Minnesota the risk is much smaller than larger metropolitan areas like New York, Chicago, or Detroit. Still, the potential risk is there. If we needed to use a gun to defend ourselves against evildoers, not neighbors knocking at our door needing food or shelter, we feel we would be an agent of God, administering justice and protection for the weak and helpless. Once we have a gun, we may be the only law enforcement in our neighborhood.

There is one other risk that sort of worries us even more.

A few paragraphs ago, we made the reference to wanting "the right to own a gun if we chose to." If things begin to "start looking ugly", by the time it starts getting that way, it may be too late. The way our courts and governments (local to federal) take away our private rights on a regular basis, we believe it would be nothing for our government, in their infinite wisdom and desire to protect us from ourselves, to make guns illegal to buy or sell, or at the very least, very difficult to obtain. By the time you might need to buy a gun, it may be too late. You may not be able to. We are not willing to take that gamble.

For those of you who already own a gun, and those planning to buy a gun, we have these two pieces of advice:

- **Safety first.** Do not even think about buying a gun unless you are ready to do all the necessary steps to learn gun safety and security. Do not buy a gun if you are not willing to make the time and money commitment to learn proper gun safety, and to provide for 100% security (to prevent unauthorized usage or accidental discharge).

As they say, guns don't kill people, people do. But sometimes they do accidentally. DO NOT allow your gun to
get into any possible situation where an accident can happen.

- **Get a shotgun.** Based on most of what we have read and
been told by gun experts, the best gun to have for personal defense is a shotgun. So if you do not have a gun
and you are considering buying a gun, check into shot-
guns.

OTHER ESSENTIALS

Food

As a matter of preparedness, you should have stored *at least* a 30-
day supply of food. We feel a three month supply is better, to insure
availability and enough to share with those who did not prepare.
Other preparedness experts such as James Talmage Stevens (*Making
the Best of Basics*) or the Urban Homemaker (*The Homemaker's
Forum*) suggest having at least one year of food just as a matter of
course, regardless of Y2K. Having a supply of food on hand at all
times is not "hoarding" as the politically correct would accuse you of,
but prudently saving from the plentiful times to provide for the un-
plentiful times. Joseph of the Old Testament provides an excellent
picture of preparing for the famine with the abundance of the harvest.
It is good to go to the Lord to discern just what level of food storage
is proper for your family and your budget.

Note: IF your electricity goes off, eat the food in the refrigerator
first. Then eat the food in the freezer. Of course, if you live in a
northerly climate, you could use a cooler set outside in the snow for
your icebox. You may want to keep it in a car or garage for protection
from animals. One winter we kept extra food outside in a cooler and
the squirrels practically chewed through the plastic cooler!

Factors to consider when planning your food storage tactics:

1. Your budget. Obviously, if you are already in the position of too
much month for the income coming in, your plans will be very mea-

ger indeed. Pray that the Lord of the Harvest would provide exactly what you would need. I have talked to mothers who have trimmed their food budget (as well as other budgets) down to where they were spending less than $200/month on food! One mother we know of does this while feeding 5 growing boys and a very tall husband. Many resources exist for trimming food and other budgets. As with many other things, you can spend as much or as little as you like. Pre-packaged, freeze dried, food supplies are extremely expensive. MRE (Meals Ready to Eat) packages for one month/family cost $1,000 to $3,000. Multiply that by the number of months you want to prepare for. However, MRE's may be the best choice for some time-strapped and storage–challenged families. On the other hand, budget conscious families prefer hundred pound bags of wheat, corn, and soybeans to provide the bulk of their storage diet. A grain mixture of 40/40/20 of the above grains provides most amino acids. More details on food storage suggestions will be later in this chapter.

2. Storage facilities. You may not have the proper space to store a lot of food. Temperature, accessibility and humidity are critical factors in maintaining adequate food storage. Keeping it in the basement will work for some but make sure humidity levels are not too high for the food. Temperature extremes are also not good for certain types of food. Storing food in a place that can range from 90+ degrees in the summer to -20 degrees in the summer can shorten the shelf life of many types of food. Lastly, make sure stored food is accessible during a crisis. Having it off site, locked in a shed, or in another location can limit your ability to utilize the food when it's needed and create an added security risk from people looking to steal your food. (Do not misunderstand us, if a hungry person or family showed up at my door, we would be happy to feed them. However, we do not believe it is wise to place temptation in a sinner's path. Psalm 1:1 "Blessed is the man that walketh not in the counsel of the ungodly, nor *standeth in the way of sinners*, nor sitteth in the seat of the scornful.") It is best to keep food in your living areas to ensure temperature, humidity and accessibility. Here are some ideas to store food in an otherwise, possibly cramped house:

• Under beds

• Move the couch out 12", place buckets on floor behind

165

couch.

- Look vertically, add as many shelves as necessary in any storage areas you have now.

- Use the buckets of food and some boards to create instant shelves for the rest of your stored food.

- If you have depth in your storage areas, think three dimensionally. Run shelves perpendicular to the wall, back to back.

- Eliminate excess clutter. Do you really need to keep every back issue of *Better Homes and Garden*? (We are using this as an example- we don't have anything personally against the magazine- except that it is full of cigarette and alcohol ads- but you get the point.) Eliminate excess clothing to a degree. You may need extra shirts, jeans, socks, underwear, etc. if you cannot wash clothes for a few weeks. We suggest getting rid of any clothing that: doesn't fit, is of poor quality, or nobody wants to wear it. Eliminate excess toys. We have packed away almost all my children's toys. All but a few Legos, trucks and blocks are now in secondary storage. Our kids have hardly noticed. Eliminate excess kitchen tools. Are there gadgets you have not used for over a year? Sell it or give it away.

- Now, a truly noble tightwad would suggest holding a garage sale to eliminate excess. But weigh the costs carefully before committing to a garage sale. We have held 2 sales in our lifetime. After the first sale, we vowed to never do it again. First, pricing everything, and then sitting out for two whole days, taking money from strangers who pawed through our personal belongings and wanted to pay us nothing for anything, and making only about $100 for all our hard work. This was neither a profitable nor edifying experience! Unfortunately, we forgot our vow and had another sale a few years later. It was a virtual re-run of the first experience. We vowed once again and we will not forget again. Garage sales are a lot of work, disrupt your life, make a big mess, (We have seven kids- we will do anything to avoid a "big mess"!) and don't always make much money. Carefully evaluate the contents of your potential sale. Do you really have enough value in items that would really sell to justify the work involved? Keep in mind most people only sell a fraction of what they intend to sell.

- We prefer to give away excess. Now THAT edifies the Kingdom of God! Just don't expect others to be blessed by you giving away shoddy or worn out items. Just throw those out. Once our church held a huge swap/sale. Everybody brought his or her unused clothing and baby items to the church. One group of women separated it by size and sex. The next day, everyone came to church, rummaged through all the piles and took home only what they could truly use. Free. Trading and sharing as the Lord leads and provides is true Christian testimony. The leftovers were given to a women's shelter. You can modify this sharing plan by setting up a bulletin board where families can post their needs or their "un-needs." In other words, right now we have an unused bunk bed. We could post a description of the item and our phone number. Therefore, someone who needs a bunk bed will call us and we can work out the details. Conversely, we could post that we need a new (used) washer or new stroller, and someone having the item could call me. For sale items could be posted as well: cars, vans, houses, rentals, tools, equipment, etc. Talk to your spouse and/or pastor about these ideas and see where the Lord will lead you.

- Consider finishing off unfinished areas of your house. Make the attic into a temperature and humidity controlled space. The long-term advantages are numerous even if you never use it for food storage again: you could turn it into a study room, computer room, sewing room, toy room or other useful designation.

- Not having storage space for stored food supplies is not an acceptable excuse. If we called and told you that we

167

would give you as many silver coins as you could store in your home, believe me, you would be motivated to find space. Consider your food storage as valuable as silver, as worthy as money in the bank. You will not regret it. Even if you have to stack food in the corner of your living room, it will be worth it if we did have serious Y2K problems.

• Get together with other families and store food together at church. Check with the church leadership on this. We only recommend offsite storage as a last resort and if the church is reasonably close (less than 5 miles).

• Last but not least, pray for wisdom and creativity. Ask the Lord to show you new ideas and creative options for carving out innovative storage solutions and overlooked spaces within your existing location.

3. Your diet. Obviously if you have special dietary needs within your family, you will need to plan your food storage needs very carefully. We have a friend with 6 children. Of the six, one is a diabetic and one has celiac disease (the inability to digest gluten) Needless to say, she is juggling some very stringent diets even on a good day! Laura (celiac disease) *cannot* have any wheat gluten in any shape or form. This eliminates most cereals, bread, many prepared sauces and condiments and many other items. Nathan (diabetes) must have his meals within a precise schedule with exact serving sizes, controlling calories and fat. Christine, the mom, must procure exact types of food and of sufficient quantities to continue to prepare the proper meals in the event of shortages of these special foods. Likewise, she must keep on hand a sufficient quantity of insulin for Nathan's daily shots. Even if you do not prepare "special" diets for "special" kids, keep in mind that your family is used to eating certain types of food daily. Unless you want a great deal of additional stress and hassle, do not plan on a food storage that would entail a completely different diet than what your family is used to. Lastly, don't forget some "fun food" items such as cookies, candy, Jiffy Pop popcorn, or soda pop. Store what you eat, eat what you store and rotate, rotate, rotate!

4. Your experience. Until embarking on Y2K preparedness planning, Tammy had never personally developed any food storage plan. We are treading this new path very carefully. Our goals are modest because we know we can only achieve what we are capable of. Do not go out and spend $1,000's on food supplies that you have no clue on how to use. Get good resources and study as much as possible before buying one single thing. Talk to experienced food storage aficionados. Start small and read #3 again.

5. The size of your family. With having seven children under 14 years old, our food storage plans will be extensive. We will need larger quantities of everything.

6. Your preparation tools. Will you even have a functioning stove? Consider your cooking options when choosing your food supplies. Other options to cooking food without natural gas and/or electricity:

- A fireplace will warm foods in a pinch but can be awkward.

- If you plan to heat the food right in the can it came in, make sure you open the can and remove the label.

- Some natural gas stoves are convertible to propane with an adapter and do not require electricity for ignition or temperature control.

- Purchase a small two burner camp stove with plenty of extra propane tanks. There are a variety of types so check out your options. Camping supply, mountaineering, or preparedness supply stores carry these types of stoves along with small ovens that sit on top of your camping stove.

- Special charcoal, sterno, alcohol, or wood chip heated cooking stoves. The Urban Homemaker recommends one called the Pyromid. It allows you to cook, bake, roast, broil, toast, grill, fry, or steam your food.

- Invest in a propane BBQ grill with a side burner. Keep it free of snow and ice through January 2000. Do not try to BBQ in the house, it is dangerous.

- Wheat grinder. Will you have electricity for that grain

mill or will you want to purchase a hand grinder?

7. Practice. Whatever alternative food storage and preparation plan you make, practice using it. Try making some meals in the fireplace. Make a "family fun night" out of the event. Try going without electricity for a weekend. The children will love it and it will help you be better prepared if it becomes necessary.

8. Store enough to share with others. If you think you'll need only a one month supply for your family, double that amount if at all financially possible, and you will have the opportunity to bless others who are unprepared. You will also want to have extra for barter or trading if it ever becomes necessary.

What do we buy for our in-home food storage?

Marilyn Moll of the Urban Homemaker/ The Homemaker's Forum provided me this check list which she adapted from *Cookin' with Home Storage* by Vicki Tate. This food storage system relies primarily on dry foods. No refrigeration is needed and most products have a 1+ year shelf life. Amend quantities based on how long you expect to use your stored food. It is the easiest and the cheapest route to go. Our Handbook was not designed to be a "Y2K cookbook." And as such, we are not going to go into actual food preparation techniques. We highly recommend getting the above–mentioned book by Vicki Tate. One good source for the food storage items listed below is:

Urban Homemaker
2527 S. Dawson Way
Aurora, CO 80014
303-750-7230
303-750-2544 fax
800-55-BREAD
(www.urbanhomemaker.com)

Suggested One Year Food Storage Guidelines[132]
Table of food items and quantities

Item	Men	Women	Child (ages 1-7)
wheat	200 lbs.	150 lbs	60-100 lbs.
other grains	150 lbs.	125 lbs.	60 lbs.
legumes	75 lbs.	50 lbs.	15-25 lbs.
sweeteners	65 lbs.	60 lbs.	40 lbs.
powdered milk	60 lbs.	60 lbs.	80 lbs.
powered eggs	2 cans	2 cans	1 cans
salt	10 lbs.	10 lbs.	2-5 lbs.
fruits	25-30 lbs.	25-30 lbs.	15 lbs.
vegetables	40-45 lbs.	40-45 lbs.	15-25 lbs.
chicken, vegetable, or beef broth			
	5 lbs.	5 lbs.	2.5 lbs.
minced onion	1 can	1 can.	5 can
cheese	2 cans	2 cans.	5 can
tomato powder	2 cans	2 cans.	5 can
oil	2 gal	2 gal	1 gal
yeast	2 lbs.	2 lbs.	2 lbs
sprouting seeds	10 lbs.	10 lbs.	2-5 lbs.

Spices and Herbs
allspice
chili powder
ground cloves
Dillweed
basil
minced garlic
onion salt
parsley
seasoned salt
bay leaves
chives
cumin
dry mustard
ginger
minced onion
oregano
sage
thyme
celery seed
cinnamon
curry powder
garlic powder
pepper
nutmeg
paprika
salt
white pepper

Baking Supplies
Almond extract
cornstarch
raisins
baking powder
dry cocoa
vanilla extract
baking soda
food coloring
instant yeast such as Saf-Yeast

Condiments
Beef broth
lemon juice
grated parmesan
Worcestershire sauce
chicken broth
mayonnaise
soy sauce
catsup
mustard
vinegar

Sprouting

Sprouting is an easy and cheap way to supplement your diet with a fresh and nutritious food. Sprouting kits are inexpensive and easy to use. You can sprout almost any kind of seed. Sprouts add a delicious crunch to any salad or sandwich.

Other methods of food storage:

- Freezing.
- Canning: Pressure or Water bath.
- Smoking.
- Salting and Brining.
- Pickling.
- Fermentation.
- Drying.
- Cellar storage.

Each method involves a certain amount of initial investment and a learning curve to become proficient. Many moms we know do a lot of canning and freezing. If you are interested in that route, most libraries or state agriculture extension offices have abundant literature and resources.

Other Food Items to Consider

- Ready–to–eat canned meats, fruits, and vegetables
- Canned juices, milk, and soups
- Vitamins
- Ready–to–eat cereals
- Vegetable oils
- High–energy foods such as peanut butter, crackers, nuts, jelly, health food bars, trail mix

Storage Tips

- Keep food in the driest and coolest spot in the house–a dark area if possible.
- Mark the date you purchased the item right on the package. We just mark the month and year: e.g. 1/99.

- Rotate your food supplies to keep it as fresh as possible.

- Refer to James Talmadge Stevens' book, *"Don't Get Caught With Your Pantry Down"* for specific food item shelf lives and preservation techiques.

- Keep food covered at all times.

- Open food boxes or cans carefully so that you can close them tightly after each use.

- Wrap cookies, and crackers in plastic bags and keep them in tight containers.

- Empty opened packages of sugar, dried fruits and nuts into screw–top jars or airtight cans to protect them from pests. (No, we don't mean your kids but rather those six–legged variety!)

- Inspect all food containers for signs of spoilage before use. [133]

OTHER ESSENTIAL SUPPLIES

Here are some other simple, practical, and easy-to-find items to purchase for some basic preparedness. Each item should be purchased in sufficient quantity to last 1-12 months, depending on your chosen level of preparedness.

CLOTHING

- Extra clean clothes (socks, underwear.)
- Warm clothing (in the north.)
- High quality outdoor wear, boots, or specially wear such as polypropylene.
- Good, heavy sleeping bags or extra heavy blankets.

COMMUNICATIONS

- Radio. We recommend purchasing a battery operated or hand crank short wave or CB radio.
- Cell phone. Digital cell phone satellites may still work when regular phone service is out.

ALTERNATIVE LIGHTING AND HEATING

- Hand crank flashlights and lamps.
- Wood for heat if you have a fireplace.
- Flashlights, lightsticks, candles, kerosene lamps. (Place flashlights in strategic places around the home.)
- Matches, lighters.
- Batteries: lots of batteries

Extra car batteries

12 Volt

9 volt

AAA

AA

C

D

PERSONAL NEEDS

- Vitamins and food supplements.
- Non–prescription drugs such as aspirin, acetaminophen, anti-diarrhea medication, antacids, laxatives, etc.
- Diaper wipes and towelettes
- Good sewing kit with needles, safety pins and thread
- Toothpaste, toothbrushes, dental floss.
- Soaps and cleansers.
- Shampoo.
- Feminine hygiene products.
- Disposable diapers and formula.
- Paper towels, paper plates, and napkins.
- Well–stocked first aid kit
- Syrup of Ipecac

TOOLS AND EQUIPMENT

- Hardware supplies such as duct tape, nails, screws, and staples.
- Fire extinguishers
- Non-electric can opener, utility knife
- Hand tools such as hammers, non-electric drills, pliers, shovels, saws, etc.
- Camp stove and fuel.
- Gas powered chain saw with extra gasoline.
- Basic sewing supplies.
- Rodent poison and bug spray.
- Signal flare
- Compass
- Large and small trash bags and ziplock bags
- Aluminum foil
- Whistle

We have tried to cover most of the things that everyone should have plenty of on hand. But many families have very special essential needs, which are uncommon or completely unneeded by most families, such as specific medicines like insulin. Other examples are:

SPECIALTY ITEMS

- Special dietary needs.
- Oxygen tanks.
- Various medical supplies.
- Contact lens and supplies
- Second, spare pair of glasses.
- Disposable contacts and contact solutions.
- Insulin
- Denture needs.
- Prescription medicines. Note: Most doctors will give you a prescription for an extended supply if you tell him why.

You are being prepared for unforeseen interruption of your supply.

BARTER ITEMS

- Ammunition
- Coffee and tea
- Sugar
- Toilet paper
- Candy
- Lighters

RESOURCES

There are resources you will want to have on hand for the coming crisis. It is important to build an emergency preparedness library. They offer specific how-tos and tips beyond the scope of this Handbook. Skills you may want these books to cover include:

TOPICS

- Setting a broken bone, delivering a baby, or cure an infection.
- Finding alternate sources of water and assuring its safety.
- Finding alternate sources of food such hunting for deer, butchering a cow, or finding edible plants in the woods.
- Learning how to dispose of human waste.
- Choosing and using stored foods.
- Consider taking a first aid or EMT class.

TITLES

Check out this web site for free **FEMA** information on disaster preparedness. www.isd.net/stobin/fema/emfdwtr.html

The Encyclopedia of Country Living by Carla Emery. This huge book, over 800 pages, is a great resource for homesteading. How to do practically anything yourself. How to birth a baby, how to raise rabbits, how to bury the dead, how to plant trees, how to buy at an auction, how to buy land. Great resource!

Making the Best of Basics- A Family Preparedness Handbook by James Talmadge Stevens. If you can only buy one book on preparedness, get this one. With detailed lists and instructions on food shelf life and recommended purchase quantities, this is an invaluable shopping tool. Also included is information on alternative heating, lighting and water sources.

Don't Get Caught With Your Pantry Down by James Talmadge Stevens. Excellent resource for tracking down preparedness items and how to plan your food storage.

The Sense of Survival by J. Allan South. Recommended by the **Urban Homemaker**, it offers comprehensive advice on preparedness. It also has extensive information on surviving a nuclear blast.

Cookin' With Home Storage by Vicki Tate. This book will help use the food you have so painstakingly stored. It is a good idea to practice cooking with stored food now.

A Year's Supply by Barry Crockett.

When There is No Doctor by David Werner.

When There is No Dentist by Murray Dickson.

Ball Blue Book THE book on canning from the Ball Canning Company.

Keeping the Harvest-Preserving Your Fruits, Vegetables, and Herbs by Nancy Chiofi and Gretchen Mead.

Most of these books will be available at our web site (www.homecomputermarket.com) or at **The Urban Homemaker's** web site. (www.urbanhomemaker.com)

Greater preparation strategies

For greater preparation strategies, you have to spend more time trying to project just how bad you think it might be, before you can do your preparing. As you consider greater preparation strategies you need to start focusing on areas outside of what would be generally thought of as basic stocking up. Greater preparation focuses on dealing with long term solutions to serious potential problems. These potential problems would be a combination of things like we have previously discussed such as no electricity, no water, no banking, no consumer goods in stores, and even no job.

Now, you may look at that previous sentence and think, "That sounds pretty much like worse case scenarios to me," but it is not. See the following section for the worst.

As you make you preparation strategy, you have to decide which of these things you want to be prepared for. The more system failures you think we will experience, the more you need to prepare. The longer you think we will experience these system failures, the more you need to prepare.

It would be very presumptuous of us to say, "You need to do this, this, and this." No one can do that because, as we have already pointed-ed out, no one is able to say with any degree of certainty exactly what will and will not fail, and for how long. All we can do is point out the possibilities and give you sound advice on how to prepare for each.

What is your "Worst Case Scenario" contingency plan

Everyone needs to have a "Plan B" or "fallback" plans if things get bad. Whatever level of preparedness you are planning now, sit down with your wife or husband and outline an additional strategy for a "what if" scenario. What if things really do fall apart, what will you do? Will you need to find somewhere else to live? What if there is no electricity, no water, no food supply, and or civil unrest (rioting, loot-ing etc.)? Find out what your church is planning based on this sce-nario. Could you be a part of a community action team? Coordinating manpower and supplying resources to serve a needy community can be a significant ministry with an eternal impact. If your church refus-es to make any such plans, get in touch with your city hall and offer your services there. If no one is planning anything, you may as well plan on getting out of town if things start looking bad. Start planning now. We are not saying it will happen. What we are saying is you need to know in advance what you will do, where will you go, and/or who will you stay with if the worst comes. Will they be prepared enough to handle your family as well? What if other families are going where you want to go? Will resources stretch out for all of you?

If you decide to move from your current location, here are some tips that Michael Hyatt gives in his book, *The Millennium Bug*:

"What should you look for in your target location? I would suggest the following:

- A small population...
- A volunteer fire department...
- Slack zoning laws...
- An armed citizenry...
- Low crime rates..."[134]

Don't be a "Wacko"

As you read this next section, please keep in mind what we see as the most important factors to consider as you examine this whole Y2K issue:

- Before everything else, know that you are a Christian. "What profiteth a man if he gain the whole world and loses his soul." We would rather have our salvation then to trade places with Bill Gates and all his wealth and power.

- Your personal spiritual growth and the spiritual growth of your family are far more important than any inconveniences you might suffer as a result of Y2K.

- Your personal witness and testimony are also more important than any inconveniences you might suffer as a result of Y2K.

As we mentioned earlier in the Handbook, over the years, as a dealer in gold, silver, and rare coins, Dan has dealt with a lot of different kinds of people. And he learned early on that some of those people bought gold and silver coins (and only coins - not bars or ingots) for some strange reasons. Let us stress that despite any "strangeness" the general population might see in these people, most of them were very nice, pleasant to be with, and very friendly. Many of them (probably most) were very sincere Christians. They had very good, sound doctrine and a good general knowledge and understanding of the Bible. Some of them were even friends of ours. But with this all said, many of these fine folks were what much of the general public would define as "wacko." We will elaborate.

Although these were very nice, friendly people, they had one of two beliefs that separated them from the rest of the general public. This separating belief was usually related to either a monetary issue or a political issue. Surprising enough, these beliefs were usually not a religious related issue. But because these people were generally also of very strong faith and religious convictions, most of the general public around them labeled them as "religious wackos", unjustly lumping in their faith as part of their strangeness.

One of the reasons why these "wackos" tended to attract this label to themselves was because they had this tendency to surround themselves with, and focus most of their lives around this one issue that was not in line with the general population's thinking and beliefs.

Giving you a little insight into these "wackos'" thinking, will help you avoid making the same mistakes that got them thrown into the wacko category. Something we hope you will avoid as you start your Y2K preparations. We feel it is very important to avoid getting labeled a wild, end of the world, and/or extremist for a very important reason.

The reason to avoid being labeled is the same reason why years ago we purposed not to follow the path of many of these people. Even if they were right about their views of personal freedom and personal rights, through their actions and beliefs, they destroyed any and all chance to witness to those around them. We purposed to place our need to have a good witness and testimony to an unsaved world above our personal "political rights and freedoms."

In the light of eternity, your witness, the opportunity to save a lost soul from going to hell, is extremely important. What could be more important? So as you go about making your preparations, do so quietly and discreetly. This doesn't mean you should not discuss this with anyone. It means you use tact and wise judgment. You can bring up the subject along with some of the more confirmable potential problems without suggesting civilization as we know it is coming to and end, or that they need to sell everything and head to the hills.

You want to be able to be open to discussing this with skeptical friends, neighbors, co-workers, relatives, or whomever, without them labeling you a "wacko." Now the day may come then they might realize all the Y2K wackos were right. But that may be too late to help them. You want to be able to keep the door of communications open between you and them, so if the situation presents itself, you will be able to give them some wise counsel as to what they need to do prepare for Y2K.

But even more important than Y2K is the Day of Judgment we will all face after death. As you keep the doors of communications open, the more and better the opportunity to witness to them about a greater, far more certain and serious crisis they will face eventually, if they die without knowing Jesus Christ as their personal Savior and Redeemer of their sins.

So if you want to be both a good neighbor and a good witness, you will want to handle all Y2K conversations carefully, tactfully, and discreetly.

Summary

In summary, as you make your plans to prepare, what ever your plans are, look at ALL the costs. Both the costs of *not* preparing, and the costs of *over* preparing. And don't just look at the financial costs. There are many other kinds of costs like time, energy, relationships, witnessing potential, and many more. Outside of your families' personal safety, your witness and personal testimony may be the greatest thing at risk here.

And don't forget to make your "Plan B" in case things do get worse than you planned and prepared for. Keep your finger in the wind, always, and don't stop gathering information.

SECTION IV

YOUR MONEY AND Y2K

CHAPTER 10

WHAT TO SELL

First and most important, have a supply of cash (Federal Reserve Note) outside of the bank. We went into more detail about cash strategies in Chapter 3. But before we go into the issue of what to sell, we want you to keep this cash reserve advice in mind as you are reading these chapters on money and investments.

Un-investing/Things to liquidate

We would say that you first need to deal with what you are ALREADY investing in before you examine the issue of what to invest in, based on the pending Y2K situation.

This advice is true of people in all income levels. Even if you do not have much money coming in right now, you may have either some savings or investments, or other things of potential cash value.

No matter what your income level, you really need to take a very serious look at all your current assets: cash, mutual funds, bonds, and especially stocks and real estate (both investment and your own personal home). You also need to seriously examine the possibility of significant liquidation of nearly all of the more esoteric investments you may own (rare coins, baseball cards, antiques, gem stones, Barbie dolls, and EVERY other collectible-including those Beanie Babies!) Also consider your age and long–term investment goals. Larry Burkett recommends that older investors, closer to retirement age, definitely look to safeguarding their investments by focusing "on asset protection as your strategy over asset growth. Protection of your principal is so much more important than maximizing profits."[135]

The Stock Market

As we pointed out in Chapter 3, we feel a stock market correction is very likely. Please re–read this section and seriously evaluate every stock or mutual fund you are currently invested in. The next time the stock market corrects downward, it is unlikely to come back as quickly as it did in 1998. The next bear market may be unbearably long and hard. Larry Burkett recommends focusing "on asset preservation rather than asset growth. I'd recommend 70-30: 70% to safer [and more liquid] assets—CDs, bonds, treasury bills—and 30% to conservative stock index funds. Buy quality—if you're going to hold corporate stocks at all, you had better hold stocks of solid, quality companies."[136]

Other areas of personal assets

Basically, except possibly your personal home, in a word: SELL. If Y2K is to have any effect on our economy, it is most likely to be recessionary. Sell now because we will probably be hitting a serious recession, and there are very few investments that actually go up in a recession. An age–old truth in investing: *You can't ever lose money by taking a profit.*

As far as selling your home, we dealt a lot with the economics of the real estate market in the "highest probability scenario" chapter. It comes down to this. Some areas of the country will get hit with dropping real estate prices harder than other areas will get hit. You have to look at just how overvalued the real estate market is in your area. Some markets could go down significantly. Other areas have not gone up so much, so they are less likely to drop. You also need to factor in the urbanization of your area. If you live in an area of very high population density, you could be at much higher risk. If there are any kind of social problems or civil unrest, they will most likely be worse the higher the population density. The more problems, the more it will affect real estate values.

We discuss the issue of civil unrest more in another section. But in summary of that section, we are basically warning people who live in areas that would be defined as the major metropolitan areas of the country. Those are the areas of highest risk. Whereas we are not telling everyone to move out to the country, those currently living in the largest metro areas need to seriously consider the risks. Personally, if we lived in Chicago, New York, Los Angeles, or one of the other really large cities, we would consider moving to a much smaller city. Now, if you feel led by God to stay, and you are really convinced it is His will, than stay. The Church would have a wonderful opportunity to minister in these areas. Otherwise, we would encourage you to seriously consider moving.

If your home is priced down towards the lower end of the real estate market, you are unlikely to gain much financially by selling now. This is especially true if you factor in the cost of rent while you are waiting for prices to fall. And the cheaper the home, the less the prices drop when the market is weak. On the other hand, going back to what we just said, if you were planning to move anyway, this could be very good timing to get a better price out of your home than if you waited.

On the other hand, if your home is in the upper price range, you are at a much greater risk to lose market value. Higher priced homes can become very difficult to sell during an economic downturn and if you lost some or all of your income, you would be in a very uncomfortable spot. Downsizing your mortgage and moving to a cheaper home may be an excellent strategy during these uncertain times. Of course, were your home were debt–free, you would only have to worry about making utility and property tax payments.

If you do decide to sell your home, the next obvious issue to be dealt with is literally, "where do you go from here?" Our advice is rent or buy a much cheaper house. Try to "sit on the sidelines" for a while and let this Y2K thing ride itself out. This advice is especially true if you have desired anyway to move to a different house, neighborhood, city, or state. Having the cash available for post–Y2K bargains could be a wonderful opportunity to purchase real estate for substantial savings.

The Realist's Golden Rule of Investing: What you need to buy goes up and what you need to sell goes down.

This is not our original idea. We do not know why it is, but it seems like those things you need to buy keep going up in price. Along with this, it seems like you always hear about these great investments after they have already gone up in value. On the flip side of this, those things we tend to need to sell seem to be down in price when we either decide to sell or need to sell. There are many factors for this, but one of the big ones is basic supply and demand. People tend to do the same things at the same time for similar reasons. Please keep this basic concept in mind as you read the next few pages.

Also keep this "disclaimer" in mind. As we have said before, we are not professional economists. Nor do we claim to have any specialized training in economics. Much of what we speculate will happen is based on a combination of studying current information available at the time of this writing, along with factoring in basic human behavior. We also researched the opinions of many well-known economists. There are certain things that all currently available evidence points to. Other things we speculate will happen are based on basic economics.

No matter what anyone says, absolutely no one knows with any high degree of certainty what will happen come January 2000. But one thing is certain; at best it will be a rough ride.

Beanie Babies and other Collectibles

We will start with Beanie Babies. Really.

But even before we get into that we will give you a real life situation that happened between Dan and two different collectors. As we explain this, please keep in mind your own collectible treasures (if you have them. We realize that not everyone has a serious case of being infected by the bite of the collector bug. And yes, we are talking to you Beanie Baby collectors and baseball card collectors and Barbie collectors too!) There were two rare coin collectors, which Dan dealt with very closely. Both were extremely passionate about their areas of interest within the field of rare coins. Both collectors' "primary collecting factor" was the esthetic beauty, and both collectors had very good eyes for picking out coins with the greatest "eye appeal" as it is refereed to in rare coins. One collector (collector A) liked to couple the beauty of the coin with the finest quality possible. The other collector (collector B) focused on pure beauty, independent of condition or rarity. (Yes, this is all relevant) Both collectors were as true of a collector as the word implies, yet neither ignored the investment aspect of their financial expenditures. In fact, collector A was very sensitive to the investment value of his purchases.

The day came when the rare coin market had skyrocketed to incredible new highs. Many of the coins these two collectors owned had doubled, tripled, and even quadrupled in price. Dan approached both collectors separately and told them the same advice. Based on his expert opinion as a professional rare coin dealer, and as a friend, he told them the market was peaking and they should consider selling their coins. Dan pointed out it may be a long time before they ever saw prices like this again and it was the best time to cash out, wait for the market to drop, and buy back in then.

Now, one thing about certain areas of coin collecting is that certain types of coins are truly so scarce or rare that a professional rare coin dealer might only see certain truly spectacular specimens once every two, three, even four or more years. These were some of the types of coins these two collectors liked. And these types of coins were not like a Mint Hank Aaron rookie card or new in the box Barbie which might be scarce and expensive, but if you had the money in hand, you can pretty much buy one any time with a bit of shopping. Some of these coins were nearly impossible to find. So this was a very hard decision these two collectors had to make.

Dan made the following challenge to each of these collectors. (This is also where all this becomes VERY relevant for all of you who know who you are and what your collector's weakness is.) "If the price of any of these coins dropped in half, will you wish you had sold it? What if it dropped even more in price?"

With that challenge, they each sold off a few things, but not the "Crown Jewels" of their collections.

So what became of all this?

The "Crown Jewel" of collector A dropped to less then one-third the value at its peak, a drop of about $20,000. Many of his other coins suffered similar drops. But it wasn't just him who saw price drops. Virtually everyone in the market saw huge price drops from the peak values. Yes, there have been a few things that actually went up from that time period, but they are the rare exception. Had he to do it all over again, he realizes he should have sold nearly everything. That

189

same money would have bought some wonderful coins at the bottom of the market (if he could have found them-but that is the risk a collector runs!)

As far as collector B goes, his story ends similar, yet different. Yes, his coins also dropped in value significantly from their highs. In fact, percentage–wise many of them dropped more. Dollar wise, he did not have as much invested, so he did not have much "at risk." The difference is that he still loves his coins passionately and truly enjoys collecting them. We don't believe he would have done too much differently, except maybe selling off a few more pieces.

So for you Beanie Baby (and other) collectors out there, our message to you is very clear: SELL, SELL, SELL!!! (We will explain more about the "whys" of selling shortly.) We hope this story will give you a little different perspective on your investment/hobby. And we issue you the same challenge Dan made these two collectors, but we will put it stronger. If you knew prices of your collectibles were going to drop possibly as low as one third or more of these current prices, would you sell them now? All right, for you out there who were not answering honestly we ask yourselves to ask look at that question again! Remember that we are predicting that some collectibles will drop TO just a fraction of their current prices (not just dropping a fraction OF their price. There is a big difference between a drop OF one-third compared to a drop TO one-third of the price.)

And now for you stubborn hold outs, who still would not sell their Beanie Babies if they knew the prices would drop drastically, we will put it another way. Don't you think that when prices drop to ridiculously low prices, you wish you had sold at the old prices that now seem like they were ridiculously high?!? And now that prices are super–low, you can buy back multiples of your initial holdings.

All right, for those last few hold outs out there, we will give you the final ultimate challenge!

This is for those of you who think you will still be glad you did not sell of those collectibles, even though you could have sold for multiples of the price you can buy them for later. How many of you are willing to live the rest of your lives with a spouse who is going to be mad at you for not selling as soon as you read about it in that radical, extremist Y2K Handbook? (This one!) Or even worse, living with the spouse who will always be saying, "I told you so" even if they don't say it verbally!

See all we are really trying to do here is promote marital harmony and remove potential sources of irritation and potential bitterness! And it will probably even be a good financial decision too!

Liquidation strategy

Here is a suggested liquidation strategy.

Step 1: Between the two of you, determine the current cash liquidation market vale of the collectibles in your family- all of them. Exclude items received as special gifts, family heirlooms, or significant sentimental value. Examine each of those on an individual basis. However, if any of them are extremely valuable, seriously consider liquidating even more so then those of lesser values.

This is another thing to remember when trying to decide what to keep and what to sell. We cannot stress this next point strongly enough! No matter how much you like an item, you can probably find something else out there that you like even more. Especially if the price of that nicer something else dropped in half! Even more so if you ended up buying that nicer something else for a lot less then you sold yours for before the prices crashed!

Step 2: After you have a realistic liquidation cash value, determine a dollar value to keep, and a dollar value to sell, best based on a percentage of the total. Our advice, if you "must" keep some, only keep 10-25% of the total dollar value and sell the rest. If (when) the market prices collapse due to Y2K related issues, you will be glad you sold what you did.

Step 3: As you are determining what to keep, do not keep ANYTHING which is easily replaceable, regardless of price. It is amazing how easy it is to buy 10, 20, or even 30 or more of a collectible item which is supposed to be scare or even rare.

Dan knows this first hand, because he has done this for a living with more than rare coins. He also owned at one time (before he gave his priorities over to God) a successful baseball card and comic book store.

There is a reason why it is easy to buy 10, 20, or even 30 or more of a collectible item which is supposed to be scarce or even rare. It is because those items tend to be advertised proportionately more in the trade shows, collector magazines, and shops that are selling them. If you really want to keep some items, keep only the truly scarcest and hardest to find. Or keep the ones which you love and enjoy the most (that first one you ever bought, the one you owned as a child, or maybe the ones which are the most enjoyable to display and look at.) With most collectibles, the only really limiting factor to acquiring them (if you know the proper trade publications to look in) is the limit of the checkbook, not the limit of availability,

If liquidating these collectibles will, in itself, be something that might potentially result in bitterness and resentment for the collector, you need to deal with that BEFORE you sell. You both need to have a oneness in your spirits about this decision. Maybe the acceptable compromise will be to liquidate a smaller percentage then originally discussed, but not as many as other people might sell.

Liquidation tips:

Having been around collectibles for many years, we learned something early on, which most collectors do not find out until too late. Although collectibles, antiques, and such are relatively easy to acquire (if you have the money), they can be extremely difficult and illiquid to sell. Most collectors have no idea just how difficult it usually is, and how much work it can be.

The buy/sell spread

We would say that learning how to sell collectibles is the first step (and one of the most important steps) in successfully investing and making true profits from collecting. No collector/investor will ever make a profit on what they collect/invest until they do learn how to sell. This is where the really rude awakening begins to happen.

The first step to selling is to learn and understand about the "buy/sell" spread before you buy (if possible.) If you already own things you need to sell, understanding the "buy/sell" spread will at least help you to get a better price, and to know when you have been made a fair offer by a potential buyer of what you want to sell.

We are going to take a side road here for a moment to explain how all this applies to Y2K. If you think you want to buy any gold or silver of any kind as part of your personal Y2K preparedness, all this is very important to know and understand in order to get the best value for your money as you make your gold or silver purchases. Independent of gold or silver, and even independent of Y2K, knowing and understanding the buy/sell spread will be helpful to you for many other uses in your day to day life when ever you need to buy or sell anything with other private parties, and sometimes even with businesses. We will get into buying and selling skills more, later in the bartering section. And now, back to our previously scheduled explanation.

"Bought right is half sold"

The second step to selling is to get a good buy at the start. In the coin business (and probably many other businesses) we had a saying, "Bought right is half sold." What this meant was that the secret to making a good sale, was to make a great purchase at the start. If you pay too much, it is harder to sell for a profit. If you pay way too much, you may never make a profit. On the other hand, if you can get a really good price, it makes it significantly easier to sell for a profit. We will show this through explaining how buy/sell spreads work.

Example:

A collector buys a collectible item for $40. The "price" or "value" of the item "goes up" to $60. For whatever reason, the collector decides to sell. But the collector can only get $30-$40 dollars for it. "But the price went up 50%," the collector exclaims! Well, yes, maybe on paper, but not in reality. Here is the reality. The $40 collectible probably traded between dealers for between $25-$30 wholesale. When the "list" price "went up" to $60, it was probably really trading at around $45-$50 wholesale. So when the collector wants to sell it, the dealer thinks, "Since I can only get $45-$50 for this if I have to sell it quick, I will try offering $35, and pay $40 only if the collector won't sell for the original $35 offer."

So the collector who's investment "went up" by 50%, ends up only breaking even. This is NOT just a theoretical example. We saw this same type of scenario played time after time after time. But usually the dollar amounts were much higher.

193

We cannot stress these next points enough.

After years and years of working with all kinds of investors and collectors, one theme seemed the most common at every end of the spectrum. When it came time to actually sell and cash out, very few made a better return of their money then simply putting the money in the bank to start with. Those were the few that did well. On the other hand, the vast majority of the people Dan dealt with lost money. Some lost a lot of money. When it came right down selling off and cashing out, most regretted the experience.

This next statement is directed specifically at any of you thinking of buying rare coins as a hedge against Y2K financial problems. (This is a little out of place, because we deal with the issue of gold and silver in another section, but it needs to be said here also. This next warning only applies in regard to collectible and investment grade coins, not to coins whose value is primarily based on the gold or silver content)

In all of Dan's years as a rare coin dealer, he saw very few coin collectors who actually managed to sell their coins for a reasonable profit. When we say collectors, we are referring to people who loved collecting coins, studied coins, and spent a lot of time trying to learn about what they were doing. But for most collectors, when it came time to sell, they made either very little profit, or lost money. As for the investors, the ones who bought rare coins strictly for profit, it was rare to see any investors who made overall good profits. And those were usually only the ones who studied the rare coin market with great intensity. The sad fact of the matter is most investors ended up losing a lot of money.

There is only one person who is assured to make a good profit in the rare coin market, the dealer who sells the coins to the investor or collector. They made the money on the mark-up (buy for $300, sell for $350), not on the coin actually going up in value.

To really drive home this point, we will give this black and white warning: except for the very few, already skilled, seasoned coin collectors and/or investors out there, we are telling every one of you: DO NOT BUY RARE COINS, FOR ANY REASON! PERIOD!

Now, as we just qualified, this "DO NOT BUY!" warning refers to coins where the value is based on rarity or collectibility. This does not refer to coins bought strictly for the gold or silver content value. We deal with that in another section.

Tips for collector/investors of all sorts (not just coins)

If you are investing for profit, make sure you know what the true *buy-sell* spreads are. Buy–sell spreads are the true prices which dealers pay when they buy items and what they sell items for. Many collectibles have no real, established buy–sell spread. This means there is no easy way to find and/or track current prices that reflect what prices dealers will buy or sell for on a regular basis. This is just the opposite of stocks. At any moment, you can get a price that you can sell a stock for immediately. Or, on the other side of the coin, you can find out at what price you can buy a stock for. And as you surely realize, those two prices are usually just a few percentages apart. Most collectibles, on the other hand, are quite the opposite. Except for a few examples of really hot collectibles, it is usually very hard to find listed prices that dealers actually pay for them. And except for those few examples of hot collectibles, most collectibles have very wide spreads between the buy and sell prices. This means that in the real world, for most collectibles, the prices need to really go up before you can even break even. As shown in our earlier example, it is not unusual for prices to go up 50% on paper, but when a collector actually tries to cash out, they end up just breaking even or making only a slight profit.

If you really feel compelled to invest in collectibles, try to stick only to things that have close buy–sell spreads, which are easy to follow.

Selling or Liquidating

When it comes time to sell, try to find other collectors to sell to. A good method is setting up a booth at the type of convention where collectors come to buy. But keep in mind these warnings: There are costs and various expenses to exhibiting at these events. These expenses will cut into any potential profits (or contribute to losses). If you are really serious about liquidating at these conventions, you will need to make sure your prices are the lowest at that show. Even then, you probably won't sell everything. If you really are serious about getting things sold, you have to forget what you paid, and focus on pricing everything lower then everyone else. Unless you can really create the impression with convention attendees that your prices are the best at that convention, they will spend their money somewhere else.

Another good place to sell is advertising in specialized trade publications that cater to other collectors like you. The same pricing rules apply as in the previous paragraph.

If you are looking for the cheapest way to sell your stuff, try having a garage sale. Another place to try sell is at a flea market. Booths are usually not too expensive and you can often find someone to split a booth with you to cut your costs in half. If you do split a booth with someone, be absolutely, positively sure that you can trust them completely.

One last warning. And we know this from personal experience. There are people who go to flea markets, garage sales, collectible conventions, and such, just for the sole purpose of looking for stuff to steal. And they usually target new exhibitors, who are less likely to be skilled in protecting their merchandise from theft. So be ready, be watching, and beware!

CHAPTER 11

WHAT TO BUY

What will the economy do?

Before you do any investing whatsoever, you first need to make an assessment of what direction you believe the economy will be going in the months ahead. Recession? Inflation? Stagflation (both inflation and recession at the same time)?

What to Invest In

Debt: The consequences and what you should do

There are three different levels of income and assets, each of which must be examined independently to determine the best approach for your family. Obviously, you would want to follow the approach that reflects your family's income and asset level. These are the three different levels of income and assets:

> More month than money
> More money than month
> More money than most

The first investment any family should make will need to be in the form of your food supply and other essential items. Each family will need to determine for themselves whether they want to have a 30-day supply, 60-day supply or whatever. Until this issue is addressed and planned for, the issue of other investments is a moot point. It should go without saying that if your family is like many which we know (and often ourselves included) many times the money runs out before the end of the month does. If you fall into this category, the issue of investing for Y2K is pretty irrelevant. You need to focus on making sure you have a good stockpile of foods and basic essentials. If things get bad, or if things are wonderful, you still need to eat, and you will still need the basics.

The only other "investment" you should be looking at, if any, would be having a supply of Federal Reserve Notes for emergency use. If you have debt (as most people do), it is important to pay off what you can. However, it more important to establish a cash reserve for at least one month of expenses. When Y2K is over, and if nothing significant comes out of it, take this extra emergency cash and put it towards any remaining debt. As mentioned previously, having a "stash of cash" is discussed elsewhere in this Handbook, so we won't go into any more detail on that issue now.

You do not actually need to purchase all these things before you deal with your other investment areas. In general, we suggest you buy the basics first, but there is no reason why you cannot be systematically acquiring the items on your basics list, while dealing with your investments. However, one thing is sure, unless you are absolutely positive you will be having a good cash flow in the coming months, you should set aside your money now to be buying your essentials.

A side note as we go through this section. Nowhere will we be suggesting or implying anyone should go into debt to do anything we recommend. In fact, we would recommend people get out of debt, wherever, however, whenever legally and morally possible. The advantages and Biblical basis of being debt–free are discussed in many other books so we will not cover it in this Handbook. We highly recommend you get Larry Burkett's book, *Using Your Money Wisely* and the Institute in Basic Life Principles' *Men's Manual II*.

Cash is king-short term.

Federal Reserve Notes are NOT real "money" but they will work best. However, Federal Reserve Notes may EVENTUALLY become worth very little someday. But we are nowhere close to that day and

Y2K will have little or no effect on Federal Reserve Notes becoming worthless.

As we discuss this idea of "Cash is King"; we are only saying this in the context of the first half or so of 2000. We are saying, "Cash is King" in terms of a short-term investment only. Something can be a fantastic short-term investment and be a rotten long-term investment. Looking towards the year 2001 or later, cash could turn out to be a horrible long-term investment. This is especially true if the Federal Reserve and Federal government try to inflate and spend their way out of a recession. If that scenario happens, cash and interest–bearing investments such as bonds will be the *last* thing you will want your assets invested in. In fact, no matter what happens as a result of Y2K; we see the long-term value of the US dollar as being very weak. But that issue, and the issue of the dollar potentially having no real inherent value (because it is a measure of debt, not of assets or tangible value), has no relationship to the short–term argument of why you will very much want to have a significant supply of good ol' green backs.

Please note, we are *not* saying that cash is likely to be a bad long-term investment. We are also not saying it will be a good long-term investment. As we have so often stated, there are just simply too many variables to make any predictions as to the value of the dollar beyond 2000 or 2001. All this goes back to our "mantra" of keeping your finger in the wind-always. And this holds true for every investment. What is good the fall of 1999 might be bad in 2000. What is good in February of 2000 might be awful by 2001.

We do not have the time here to go into any greater detail as to why the U.S. dollar has no real value and why it a very bad long term investment. And that issue has little to do with the focus of this book. So we will leave this topic as something for you to go out and independently research if you desire. If you just ask around in your own circle of personal friends and acquaintances, we are sure you will find someone knowledgeable in hard assets who can recommend some additional reading material on the subject of hard assets and the intrinsic worthlessness of the U.S. dollar.

Just remember, any information promoting the advantages of hard assets is usually written with a number of strong biases. Most of the people who write this material, and promote it, are in some way usually connected to the *sale* of hard assets. Even for the most mature Christian, it is hard not to have an unfair bias towards something you sell and make a profit from.

Another factor is that most of the hard asset mentality comes from an economic reasoning which is either "pre-Y2K" aware, or from a flawed (in our opinion) viewpoint. Gold and silver may someday turn out to be a fantastic investment. But that day is more likely to be sometime way down the road. Possibly as far as 5 to 10 years down the road. Until then, be very aware of what is happening now and in the next 6 to 18 months. And don't forget to keep your finger in the wind and be ready to change investment strategies when the economic winds begin to change.

CHAPTER 12

WHAT NOT TO BUY

(Author's note: We made this section drawn out and detailed because there is tons of information, hype, and pressure to buy gold and silver. The vast majority is very one sided, without explaining the various negative aspects of buying gold and silver. There is almost no information out there that explains why you might not want to buy gold and silver. So although this section may seem a little longwinded for some, we really felt we needed to say what we did to help counter balance all the hype you will hear telling you to buy gold and silver as you read other materials on Y2K.)

It should go without saying that you will not want to invest in any of the things we are suggesting that you sell. But it is important enough that we are just clarifying the issue.

"The Case Against Gold or Silver"

First, what does the Bible say about gold and silver?

What does Jesus Christ say about this subject? His words are "Lay not up for yourselves treasures upon earth, where moth and rust doth corrupt, and where thieves break through and steal. But lay up for yourselves treasures in heaven, where neither moth nor rust doth corrupt, and where thieves do not break through nor steal. For where your treasure is, there will your heart be also."

The Bible cautions against trusting in gold and silver. If this really is the judgment of God, a stockpile of gold or silver will be of no help to us. Zephaniah 1:18 says "Neither their silver nor their gold shall be able to deliver them in the day of the LORD'S wrath; but the whole land shall be devoured by the fire of his jealousy: for he shall make even a speedy riddance of all them that dwell in the land."

James 4:1 Go to now, ye rich men, weep and howl for your miseries that shall come upon you. Your riches are corrupted, and your garments are moth-eaten. Your gold and silver is cankered; and the rust of them shall be a witness against you, and shall eat your flesh as it were fire. Ye have heaped treasure together for the last days.

If you are planning on storing up gold or silver from a financial standpoint, read on. You will see from Dan's coin dealing expertise why gold and silver are a very bad financial investment. If you are looking at gold or silver from an investment stand point, seriously consider what we have to say here. Read the investment chapter for our recommended suggested investments.

Keep in mind, we are not Certified Financial Planners. These are our opinions, as best we can conclude. Our best advice is invest with prayer first, and make sure you have the agreement and "blessing" of your spouse before you may ANY investments. If your spouse is not of a like mind and of 100% agreement, then do not do it.

If you are considering gold or silver as the new form of money for the post-Y2K collapse, you had better start re-considering that thinking right now, as we will also cover here. Greed is the bait that is driving most to look towards gold and silver. And that greed will snare you right into Satan's trap. Do not let greed, fear, or anxiety cloud your judgment and cause you to make a bad decision you would otherwise never make.

We will challenge you with these questions that goes back to the main theme of our Handbook. As you look at your own preparations for the next year, century, and millennium; what spiritual preparations and changes would God have you make? What is more important, getting your finances set for 2000 and beyond, or getting your spiritual life in order? If God could direct your planning, what would He be changing in your life? What changes do you need to do in your life and your family's life to prepare for not just Y2K, but preparing spiritually for the rest of your lives?

Disagreement with others

On January 12, 1999, Michael S. Hyatt (author of *The Millennium Bug*) was featured on James Dobson's Focus on the Family national radio show. His comments on gold and silver as a Y2K investment were typical of those we have heard from so many other Y2K speakers and authors. He was discussing what to do with your investments. He was recommending getting out of the stock market (which we agree with) and then suggested where to put the money. One suggestion was putting some money into bonds (which we also agree with). He followed this with the comment, "If you really want to get defensive and you have a low tolerance of risk, then you can move into silver and gold and some things like that. I think at the very least people want to have some emergency cash on hand."

We agree with the cash on–hand part, but we cannot disagree strongly enough on his assessment of gold and silver. For starters, it is our expert opinion that you should not buy any gold or silver. But we have an even greater disagreement on this next point. We do not know how much dealing he has had in buying and selling gold and silver (we speculate little or no experience) but we can tell you as a historical fact that investing in gold or silver is anything but "defensive." Historically, gold and silver have been a high–risk investment. Our own experience is that MOST of the people we have dealt with have either made very little money in gold and silver, or they lost money. And we see nothing, in our opinion, that would change that picture for the year 2000.

Background

In the late 1970's and early 1980's, inflation was up and the dollar was down. There was some degree of panic and fear over the declining dollar. And at that time, many of those fears had some valid reasons behind them. Then, along came President Ronald Reagan and "Reaganomics." At first, the term "Reaganomics" was meant as an insult. But, as they say, the proof is in the pudding. And Reaganomics turned around the economy and the value of the dollar.

The point of all this? With the change in the economy and the change in the value of the dollar, there was also a change in the philosophy behind buying gold and silver. The economic conditions and government policies which lead to the justification of buying gold and silver were now gone.

Economics

The argument for buying gold and silver lies in one key factor: the decline in the value of the U.S. dollar. If the dollar stays strong, there is no case for holding gold and silver. We will tie all this into the dollar and Y2K.

There are two perspectives on this key factor. If the dollar does become significantly devalued, then gold and silver *do* become prime candidates for a replacement. Gold and silver will be highly sought after and desired as a store of value and wealth.

Second, if the dollar does become significantly devalued, people will seek an alternative to the U.S. dollar as a means of exchange for services and products. This is supposedly the key factor to why many authors are suggesting buying gold and silver coins as part of your getting ready for Y2K.

Beware of the snake oil salesman; he has a mean bite!

Due to the hysteria caused by Y2K, lots of hucksters and snake oil salesmen have deceived enough people to believe that these are the only things that will have any value after 1/1/2000 will be old 90% silver coins and old U.S. gold pieces. All this has cause an incredible large, false demand for these coins. It all goes back to the first simple, fundamental law of economics: supply and demand. The supply is basically fixed. So value goes up and down based on the demand. Normally, supply slightly exceeds demand. But due to an over abundance of hype, hysteria, and misinformation, there is this temporary, abnormally huge demand that has created these temporary exorbitant premiums.

Ladies and Gentlemen, we are going to let you in on a very simple little secret. About the only real reason these hucksters are telling everyone to buy these grossly overprices coins is to fatten their own pockets! They have no real interest in your best interest. They really don't want to help you. Their real goal is to put your hard-earned money in their bank accounts. In the process, these scoundrels will tell you whatever twisted lie, distortion, and misrepresentation that they can come up with (without getting thrown in jail) to drive enough fear into you to do something foolish that you would never otherwise intentionally do.

Folks, if we sound just a little bit cynical here, it is only because we are. We are just fed up with the kinds of lies and half-truths we see coming from these low lifes! It just makes us sick! We just wish that we could sit down with each and every one of you and expose the lies that you are hearing, to clearly show you the specific misrepresentations in the so-called newsletters some of you are reading. Many of these things are not newsletters. They are just cleverly disguised advertisements for their products, and some of you are actually paying subscription fees for these commercials! They are not going to tell you the good new. They are not going to tell you all the great success stories that are happening. They are not going to tell you about all the banks and power companies that are saying, "we are ready for Y2K right now!" Talk about viewing the world through rose colored glasses, these people want to see the world through "red tinted" glasses, only seeing the worst that they want you to see.

Start looking for the success stories. Sure Y2K is going to cause some problems. But guess what, there are going to be a lot of shocked people when we get to the year 2001 and everyone looks back on 2000 and sees how little trouble it really was compared to what all the doom and gloom-sayers were predicting. Those folk are going to have a lot of egg on their faces and they are going to have a lot of apologizing to do. We believe that those spewing forth the prepare-for-the-worst messages will need to be held accountable for all the hysteria they preached and all the bad advice they told folks to do.

Now, if you will excuse us, we need to get down from this soapbox and get back to our previously scheduled topic.

Fast forward to 1/4/2000. When everyone realizes the banks are fine and all the stores can take your checks or credit cards, nobody, but *nobody*, will be willing to buy or take in trade any of these ridiculously overpriced old U.S. gold coins! Prices will plummet and the unfortunate people who bought them will be left holding the "golden" bag. Actually, the price plummet is likely to start sometime in late 1999, as people start to realize the doom-and-gloom, end–of–the world crowd is starting to look more and more wrong, and that Y2K will not be the end of the world as we know it. When people start trying to actually sell off some of these coins, price drops will get even steep. Our prediction is that by January or February of 2000, these types of gold coins will only be selling at around the melt–value with absolutely no premiums whatsoever.

This is our advice to those of you who have already been the unfortunate victims of all this misinformation and half-truths. If you do own any old U.S. gold coins, you should sell them right now. Do not wait until next month. Get it done now, while you can still take advantage of the high price premiums. If you still want to own some gold as an investment, do a swap of your gold coins for American Gold Eagles or Maple Leafs. Any dealer would be happy to help you (just do some shopping around for the best price on your coins first. If you are reading this "after the effect" when all the hype has died down, we would still suggest you sell them off. Prices will not be going up until gold goes up, or the snake oil salesman crowd can find another end of the world crisis to promote to fatten their wallets.

While on this subject we would like to dispel two other big arguments being used by some of these promoters of hysteria. These hucksters claim the U.S. government will be confiscating all of our gold except, of course, for the types of old U.S. gold coins these promoters are selling. Their argument here is that in 1933, the federal government confiscated all gold, except for jewelry and collectible (numismatic) coins. So, according to these promoters, the economy will fail, the banking system will not function, and the value of the dollar will collapse. Then the federal government will illegalize private ownership of gold again and come into our homes and confiscate it all. All this happening, of course, while the federal government supposedly is not able to function because of Y2K problems. But, these promoters claim, the federal government will be able function *enough* to illegalize gold and come in and take it away from us. Come on already! Or as our children would say, "Give me a break!"

The best way we can describe this fallacious argument is "bogus!" This argument just does not hold up when you really examine it. Without going into a whole chapter as to why this is wrong (and we could) we will try to summarize in two simple paragraphs. For starters, this whole argument is based on the speculation and prediction that all the catastrophic events that we just mention will actually happen. Unless nearly all those things actually happen, with the last one being our government confiscating our gold (only *after* the others have occurred or started to occur) is there any reason whatsoever that "collectible" numismatic gold will be any better than any other well recognized form of gold. (Our argument is that old numismatic gold is actually *worse* to own for an investment!)

The other argument for the government taking all the gold is, "They did it before and they will do it again!" We are sorry if we offend anyone with our response, but this argument is so stupid that it is ridiculous! That logic would also imply that the government could also bring back slavery and take away women's right to vote! But there is another reason that is more specific to the gold issue. If you go back the "why" our government confiscated and illegalized gold back in 1933, you will understand that it was part of a very complex plan to eliminate the direct tie that existed then between gold and the U.S. government issued, gold *backed* dollar. Back then, a real dollar was worth a fixed weight of gold. The government wanted to break that tie. Amongst the other reasons was that the government wanted to be able to deficit–spend and to be able to inflate its way out of the depression. Since the dollars is no longer tied to gold and they deficit–spend all they want, the reasons the government banned gold ownership once do not apply anymore. End of issue.

Fear, hype, and hysteria are great for selling newspapers, magazines, newsletters, and all kinds of preparedness related items like gold, silver, and food storage stuff. Good news doesn't sell, so you are not going to hear it. You aren't going to hear many of the success stories from them.

Let's assume for just a moment that the worst hit. The economy collapses and the federal government decides that in order to save the nation, *"for the good of the children"* it needs to ban gold and confiscate it all. Just because they did not take collector coins in 1933, what reason would they not *also* take collector gold this time, if they are taking all the other gold? What do these bullion dealers know that they can make the predictions what the government will want to confiscate and what they will *allow* the public to keep? Remember, after all, they will be doing it *"for the children!"* Who says they are going to stop at just gold bullion? Who says they will *not* go after call common, non-rare dated collector gold like the old $20 Gold Liberties? Who's to say which coins will be safe and which coins will not be safe from a law that does not even exist yet? Who is to say the Feds won't decide they want it all, *including* all the silver too? Us, that's who. It just ain't gonna happen that way. The government won't be confiscating anyone's gold or silver and that's that!

"The Article"

At this point, we would now like to bring your attention to what we believe is an extremely critical flaw in the logic of this line of thinking. And this brings us to the article that caused us to lose our faith in gold and silver as a replacement for dollars as the means of exchange of value. Or, put another way, why we no longer thought gold or silver would work as a successful means of barter.

In this article, the author set down a perspective on monetary exchange that made it clear why gold and silver would be a very inefficient barter substitute for dollars. He put it in such plain and simple terms that we were amazed we never saw it before. All he did was explain the logistics of how "value" (dollars) is exchanged for goods or services. Then he showed how this would no longer work with gold or silver. It was so simple, yet so accurate.

We are going to explain this all basically in the same manner he explained it. But first, before we do, we need to ask you to empty your mind for a moment. Empty your mind of all preconceived notions and reasoning why you might possibly need gold or silver.

Thank you. Now, focus on this example of how "money" is typically handled in our society today.

Follow the Money Trail and understand the form "money" takes as it changes hands

You get paid from your employer in the form of a printed piece of paper called a check, or the funds are electronically deposited in your bank account and you get some sort of "receipt" or check stub showing how much the government stole, oops, we mean taxed you. If you get a check, you deposit it in the bank manually by handing it over to them and they credit your account.

Now that you have control over this money in the form of a bank balance representing a specific amount of "dollars," you can spend it. So just how do you spend it? Some of it was whisked away before you even "saw" it, payroll deductions for insurance or such. Some of it will be whisked away on a set monthly interval through automatic electronic payments. So far all the spending has been automated, and you haven't needed to do one thing (except actually earn the money)!

We need to take a slight detour in our example here for just a moment. We do understand and appreciate the fact that it is because of things like this automation, and the very sensitive balance it rests upon, that is causing many to panic. Yes, potential Y2K problems could cause this fine balance to break. But we are seeing too much evidence that these sorts of potential problems are in fact being seriously worked on and (despite what some of the more alarmists are saying) will get fixed by December 31, 1999.

Back to our regularly scheduled example.

With the money left in your checking account, you sit down and pay some bills. Don't forget to pay those credit cards that you use for convenience purposes, or maybe to earn frequent flier miles. (We trust our readers do not use credit cards as a means of financing to borrow money for things they could not otherwise afford to purchase!) You make sure you pay the balance in full on each and every one so you do not have to pay a cent in interest.

With all the bills paid, now you can do some shopping. By the way, how much actual cash have you seen trade hands so far? None. Zero. Not a cent. Maybe when you are out shopping, you would handle some cash. At the fast food restaurant. To buy a newspaper. At a vending machine for some flavored carbonated sugar water. Sure, you go to an ATM or a real live teller to withdraw some money, but what actual percentage is that of what you spend? Very small.

"Why barter won't work."

(Learn and understand why barter won't work and you will be far on the way to being able to successfully barter when you need to!) The reasons why barter does not work show the great gaping hole in the entire argument for needing gold and silver to barter with. The critical flaw in this thinking stems from the idea that "barter and usage of gold and silver is how trade was done for thousands of years and we will just have to learn how to do it again."

Wrong. And let us count the ways this is wrong.

Reason # 1. Bartering skills are learned, not born.

Before the common use of stable currencies, everyone learned the skills and fine art of both negotiating and bartering. Small children learned it standing beside Mom and Dad in the marketplace, at a farm, or anywhere else trading needed to be done. Everyone practiced it practically every day.

Fast forward to today. How many of us were ever trained as children by our parents in the art of negotiating? Very few. The need to "haggle" just doesn't exist like it used to. Quite the opposite. The ability to "haggle" does not even exist in most cases. Just like our previous example of trying to use silver coins at Wal-Mart, the gas station, or to pay your bills, so too bartering does not work. Ever try to barter with a barcode–scanner at the check out? Won't happen. We will give you the fact that every now and then you will find a flawed item, and ask for a discount from the store manager. But really, that hardly counts. (Just try doing that with the electric company!)

We don't think we need to go into any more examples of the logistics that make bartering so impractical for our day and age. The following is the critical factor which makes this so.

Reason #2. You need to deal with the source or the principle party.

In days gone by, whenever people needed any goods or services, they dealt directly with the one who made it, grew it, or did it. Or they dealt with someone who was directly under the one with the goods or services. So the logistics and negotiating existed to facilitate negotiating and bartering. Such logistics no longer exist. Add to this the fact that for hundreds of years, most cultures totally lacked any true form of monetary exchange, with the exception of a unit of weight of gold or silver. Sometimes this was in the form of a coin. Sometimes not. So it was either learn the skill of negotiating and bartering, or you could not function. Today it is just not the same. Today, bartering is essentially impossible. Just try negotiating with the price scanner the next time you check out at the store. Or when the clerk asks, "That will be $52.47. Cash, check, or charge?" just try offering an even $50.00! Even if you try to speak with a manager, you most likely will be unsuccessful. Prices are usually set by a team of faceless, nameless accountants located hundreds of miles away.

Reason #3. You need to understand price vs. value.

This is somewhat related to what would also be thought of as being a shrewd consumer, the ability to understand the difference between price and value. The obvious is when a box of cereal costs more than a smaller box of the same kind, but the higher priced box is cheaper per ounce. So it is a better value. This is easy when dealing in fixed prices and fixed measurement units (ounces, pounds, gallons, feet, etc.).

This becomes significantly harder when dealing with items that do not have pre-fixed prices. Ever try to negotiate to get a higher price for a car you were selling? Sometimes it works, sometimes it doesn't. For more on this, see the next reason.

Reason #4. You need to establish an agreed upon value.

Fast forward to January or February 2000. You find a private party (not a store) who has something you want. Let's say it is a case of toilet paper. You figure in your head, "Let's see, 25 cents times 64 rolls, that comes out to $16.00 in 1999 prices." That was the easy part.

He says "I want a buck a roll, or $50.00 for the whole case."

Now comes the hard part. Maybe he would rather take silver, maybe not. If he would rather have silver than cash, you can get a better deal or a better "value" than if using only cash. Or the opposite could be true. And this is a critical point. Everyone understands dollars and the perceived value dollars basically represent. In the example, you see the value in terms of 1999 dollar values. He sees it in Y2K dollar values. Despite the differences, most people are able to re–adjust their dollar value perception relative to the goods offered. (For example, paying a dollar for a can of pop from a vending machine on a hot summer day versus the same thing in the grocery store for half the price.)

Now, change the equation. Same toilet paper as mentioned, but now the medium of exchange is silver coins. How much in silver coins per roll of toilet paper? Not only does this cause a weak link in the chain of a successful transaction; it can result in at least four (4) weak links.

Weak link A. How will you know what is value of your silver coins. You know you paid six Federal Reserve Notes for every one dollar of silver coins. But that was 1999. How many Federal Reserve Notes is that one dollar of silver coins worth now? If you do not know, it is very hard to do any successful bartering with your silver coins. (We will tell you the two best ways to know the proper value of silver coins, at any time or any price - later).

Now assume that you know what the "value" is of your silver coins, but the person you want to trade with does not know the "value."

Here we have weak link B. The only way that person will be willing to go through with the trade is if they are convinced that they are getting a value in silver coins of at least what they can easily be traded off for in equal (or better) value of the toilet paper they are trading for. (Yes, that last sentence was a mouth full! You might want to reread it!)

Weak link C. Some people simply will not be willing to take silver coins because they just cannot get a grasp on how to know the "value" of it.

Any of these three weak links can cause a barter transaction to either fail or be very unsuccessful for you. (You get "the short end of the stick" in the trade). And of course there are so many other reasons barter might not work, or work to your disadvantage.

Weak link D. Baby Boomers and Generation X. This is the fourth reason why trying to use silver coins in a barter situation might fail. A very significant percentage of our population is too young to have known the days of real silver coins circulating in change. Only the older part of the Baby Boom generation can remember the days of finding a silver dime in change and knowing it was really worth 15 cents, or a quarter, or what ever the going rate of the day was. They saw and knew that silver coins had a greater value, even if they did not know exactly how much greater value. For a very significant percentage of our population, a dime is a dime. And the concept of a silver dime being worth 4 or 5 or 6 or whatever times the value of a nickel and copper-sandwiched dime is something they cannot grasp. For many of these people, good old greenbacks will make a lot more sense to them.

Now we go from our sub-points of specifically why silver and gold will not always work for bartering, back to our main topic of "why barter won't work."

Reason #5 "see above."

That's right, look back to what we said in reason #4 "Establishing Value." Take anything else you want to use for trading, and substitute that thing in place of silver in the reason #4 of establishing value. Any of those same weak links could cause any potential barter to fail. We will review them again, in summary. As you read the list, think of other barter–able items instead of silver.

1. You need to know the real "value."

2. The person you are trading with needs to know the real value.

3. Some people just cannot get the hang of placing a non-dollar value of something. The skill or ability is just not there.

4. We have at least a generation or more of people who have never really been taught the skills to assign value in a trade to things like silver.

Reason #6. "Fear"

Just as some people can speak in front of a crowd, others are terrified at even the thought. In the same manner, various fears can have the same effect on negotiating and bartering. Here are some of the reasons, to list just a few:

• Fear of speaking to a stranger.

• Fear of failure to make a successful trade.

• Fear of making a really bad trade.

• Fear of sounding dumb.

• Fear of making a fool out of yourself.

• Fear of becoming embarrassed.

• Fear of (insert yours here!).

The key is negotiating.

Notice how we keep using the terms negotiating and bartering. It is the "n" word that is often missing in the equation of those advocating use of gold and silver for barter. The true key and secret of successful barter is when you have completed the exchange and you have something of equal or higher value, or of greater need, after the trade. The ultimate successful barter is where both parties walk away with something of higher value or greater need.

How can this be? How can you gain unless someone loses? Easy. I need a horse; you need a cow. I have a cow; you have a horse. Simple and easy. The value I place on something is going to be different from the value you place on something. That is why the prices of things in hotel vending machines are so much higher than store prices. They know that anyone interested in it has a higher value and need of it, since they are without it and they need it now. The real secret to being truly successful in a barter transaction is to find something of yours that the other party values more than you value it. And visa-versa. Trading for something of theirs that you value (or need) more than they value it. Sometimes the cost of convenience is very high as well.

Bartering truly can be a win-win situation. And when it is, you will find future barters with that person will be easier.

Fast forward from the days of horse trading to corporate America. Ask any sales manager about the negotiating ability of the average person and most will tell you that most people basically have poor negotiating skills. (This is not true for all cultures. Many cultures still teach their children the skill and art of negotiating. Just ask anyone who has conducted many international negotiations.) Most sales organizations spend considerable sums of money on teaching negotiating skills (which is basically a very sophisticated form of bartering-trying to come to an agreement on how much goods or services to trade for something else, nearly always some form of money.)

Reason #7. "Profit and Greed"

To put it bluntly, there are a lot of people out there who simply will not do a barter, trade, or cash sale for that matter, unless they are getting an increase in their value in the process. For some, unless they can make a significant increase of value, they simply will not trade. The idea of a fair and even exchange of values is something that goes against their nature. This is a major stumbling block to many trades.

Sometimes, the perceived value of an item can be misinterpreted as greed.

Let's go back to the horse and cow example. I might need a horse; you might need a cow. I have a cow; you have a horse. So does this make an even trade? Maybe, maybe not. Last week my neighbor traded his cow for a horse, three chickens, and a goat. On the other hand, your brother-in-law traded his horse for a cow and a calf. Consequently, you sincerely think you should get more than just a cow for your horse. And I sincerely think you should throw something in with the horse, to make the deal fair, otherwise I think I am getting cheated.

So did the trade work? We don't know; we just made it up! But you see the point. Sincere differences in perceived values can result in misunderstandings. Thinking that the other person is being greedy and trying to get the best of the deal can kill it. When actually it's just the different perceptions of the value of a horse vs. a cow! When both parties can understand this, it helps negotiations significantly and it is much easier to work out the perceived value differences. Of course sometimes the perceived values of what is to be traded is seen just the opposite. What I have to trade may have a greater perceived value to you than it does to me. And visa-versa, what you have to trade may have greater value to me then the value it has to you. Obviously, this situation allows for a barter type of trade to be much more likely to go though successfully.

Reason #8. "It's worth more if you own it."

In the coin business, there is a saying that goes "ownership is worth a point." Unless you understand coins, their grading, and their pricing, this expression is meaningless. But we are going to explain it to you because, like it or not "ownership is always worth a point." (Not only that, but understanding this will help you get a better idea of why you do not want to ever invest in rare coins!)

The value of a coin is based on four factors:

1. Precious metal value of the gold or silver weight

2. Demand

3. Rarity

4. Condition

For this example, factor #1, precious metal value is irrelevant. If a coin is truly "collectable," its value will far exceed the gold or silver value. The actual value of the silver or gold content is negligible relative to the numismatic or collector value. The obvious is that the nicer the coin (better condition) the more desirable it is, so it is more valuable. In the same manner, the rarer a coin is, the greater its value. Whereas demand has a significant affect on the value of a coin, it is also not relevant to this explanation. The same with rarity. Which leaves us with condition.

As you have read this whole section on bartering, some of you will think, "well that really didn't tell me much I didn't already know. This is still okay. Everyone knows that the more demand for something, the higher the price the seller can get for it. This is a general rule for anything of fixed availability. (For things that have an open ended availability, like computers, greater demand equals larger sales volumes equals larger units sold equals lower capital costs per unit made equals greater profits equals more money for research and better manufacturing methods which ultimately results in lower costs.) When there are less of an item than there are people who want it, prices go up. Think Beanie Babies. Unlike profits in a company affecting the value, nothing changes with a Beanie Baby except scarcity and demand. As demand changes, so does value. We hope we at least clarified some issues for you and at least gave you some kind of additional insights to make you a better negotiator or barterer (is that a word?)

If this is something you are already completely familiar with, this means you probably have that natural knack for negotiating and bartering, which many people (probably most) just don't have. For some of you, the whole concept of negotiating or bartering is just petrifying. Sort of in the same way some people can stand up and speak in front of an audience and others can't. So for those of you who struggle with how to negotiate and barter, we hope this has helped. It is a lot easier to "play the game" when you understand it, its steps, and its unwritten "rules!"

Going back to the original premise of "why barter won't work." If you understand the reasons why bartering fails, it will bring you a long, long way towards actually having successful trades and being a better negotiator. There is a reason why we approached the whole bartering issue from the negative side instead of the "how-to" side. Bartering is hard. And even harder is getting a really good "value." This is the real world. Most people you might barter with are less interested in what would be considered an even trade. And even fewer still would be willing to barter, just to help you out. We felt that if we prepared you for the real world of bartering, you would be in a better position to know what to expect, how to do it, and how to be successful at it. We put this section about bartering in the gold and silver section, because that is generally one of the primary reasons other people are recommending we purchase gold and silver to use for bartering.

CHAPTER 13

A CASE FOR
GOLD AND SILVER:
THE SCENARIOS

(Authors' note: If you have no interest in or plans to buy any gold or silver, do yourself a favor and just skip to the next chapter. If you think you might to buy some gold or silver at some time after this whole Y2K thing, you can always pick this Handbook up again, and read it all then. We promise, books will not be affected by Y2K, as long as you have some light to read them by :-))

We realize that for whatever reason, despite our recommendation not to buy gold or silver, some people will still want to buy some. So we have included here a section about actually buying gold or silver. Now, we understand it may seem a little confusing. "Why would the authors put a section about BUYING gold and silver in a Handbook where they recommend *against* buying gold or silver?" So why are we including these two seemingly contradictory sections? Two reasons.

- We are not saying there would never be a need to buy gold and silver. As such, we feel only appropriate that we give a balanced message. We want to lay–out what we believe would be a scenario where it probably would be wise to own some.

- For those of you who still desire to buy some gold or silver, we feel it would be wrong for us to not give you wise

219

counsel as to how best to buy it if you do buy some. With Dan's many years of personally buying and selling millions of dollars of gold and silver, he does have a significant amount of expertise that can be of great use and value to anyone wanting to buy their own gold or silver. So we felt it was very prudent we at least help you to make the best purchases possible, if you do decide to buy to buy some.

We can only see two possible scenarios where you would need gold and silver coins for barter. The first of these is in the case of total bank failure. But not bank failures in the sense of banks going under. No doubt some banks will fail as a result of the economic consequences of Y2K. The way we mean bank failures is in the context of our whole banking and monetary system "breaking." We mean "breaking" in the context of no longer working or being broken. In the past, banks failing or going under were a result of it financially going bankrupt. When a bank goes bankrupt it fails or "goes under." We do not mean this type of bank failure. (We know that the way we word this seems a bit awkward-saying "what we mean" and what we don't mean. But it is very important that you do not misconstrue what we are trying to say.) The failure we are looking at, and concerned with here, is the failure or breaking of the "system" which banking operates on, any bank and every bank. More on this in a moment.

If a specific bank "goes under" or "fails," other banks step in and take over. If need be, the government puts in money so that all the people with savings and checking accounts get back all their money on account (up to $100,000 per person.) Business continues and life and banking go on with very little impact on anyone but the investors of the stock in that bank. This is not the kind of bank failure that would drive us to gold or silver. With the type of banking regulations we have now, and with the thriving, growing economy, bank failures have been fairly uncommon as of recently.

Yes, due to the economic conditions that Y2K will bring, there could be a sharp increase in the number of banks failing and getting "absorbed" or merged into larger, stronger banks. The primary cause of these sort of bank failures will be due to the bankruptcies of many businesses that have taken out loans which they will not be able to pay off. But despite large numbers of bank failures, banks should be open and business should be able to be conducted in the form of dollars or Federal Reserve Notes.

Please keep in mind that even if we do see a total collapse of the banking system that does not mean that dollar bills will become worthless. In fact, as we have already discussed, it is our opinion that in the event of a total collapse of the banking system, actual cash or currency (dollar bills) will actually significantly go up in relative value. Even doomsday prophets such as Gary North are recommending cash as your primary hold of wealth. He forecasts the value of cash to appreciate up to 600%. Probably even initially go up in relative value greater than gold or silver would go up.

The second case for the need of gold or silver is in the total collapse of either the US dollar, and/or the collapse of the Federal government that backs that dollar. In a sentence, for the reasons outlined so much already in this Handbook, we do not think either of which will happen. We realize there are some authors on the subject of Y2K who believe Y2K related problems will cause the collapse or near collapse of either the value of the US dollar and/or the Federal dollar. To this, we simply say we believe these authors are wrong. We completely disagree with their conclusions.

A longer term look

Now this situation could end up resulting in high inflation rates 6-18 months after the government starts pumping out massive amounts of funds to prop up the banking system. So there is a good argument for gold and silver as a hedge against long-term inflation and the lowering of the dollar's value. But this is more related to buying gold and silver as an investment, a storage of wealth or value. And again, this is in terms of long term. However, mostly the call for buying gold and silver is based on the short-term demand for it as a barter tool. And as we have previously discussed, we believe this rationale is flawed and in error.

A need for gold and silver

As long as banks are able to manage the electronic flow between them and Federal Reserve banks, there is no need for a substitute for bartering. But if banks cannot successfully conduct electronic transactions for any length of time (such as perhaps more than a week or two) then it is likely to become a much larger problem. What will exacerbate the obvious problem of banks being unable to conduct transactions, is that everyone will lose faith in the banking system. Initially, this would probably cause a massive shortage of currency and massive shrinking of the money supply, actually resulting in a deflationary effect as opposed to the inflationary effect many would think would result.

Eventually, as the government pumps more and more actual printed money into the system, there would likely be an eventual inflationary effect. This inflation could seriously drive up the prices of gold and silver. But that is a more long–term effect. As we look at this scenario, we see it having only a very, very small probability of happening. Each month that passes the picture looks better and better for the banking industry and the stock market to be able to successfully conduct business on Tuesday, January 4th, 2000. (Monday will be a holiday day off for the government, banks, the Stock Exchange, and many businesses.)

Going back to the earlier reference of total bank failure: If, due to whatever cause, almost all banks are not able to function logistically then we have a serious, serious problem. Cash will still be able to be used, and will still be in great demand, as we discuss in this Handbook. But this is a situation where gold and silver would also come into very great demand, as we just elaborated.

Wild cards

Here are the two biggest wild cards to all this: electricity and massive bank runs due to panic.

If we do have massive failures in the power grid, then nearly everything fails. Due to reasons we have already covered, if we lose power, every sector of our society will fail except for the absolutely highest prepared. And even they risk failure due to supply chain problem. But again, as we look at this scenario, we see it having a very, very small probability of happening. Brown outs, possibly. Temporary outages, maybe. But a total failure of the power grid? No. They may have problems figuring out how to bill you, due to their own Y2K problems, but we should have power.

The second wild card: panic. We have already covered the general nature of panic. Specifically, as it relates to Y2K and the banking industry, panic is probably the most likely possibility of massive bank failures, although we believe this is very unlikely.

The bottom line to all this is as long as banks are operational, the general public and businesses will choose the dollar over gold and silver. If banks cannot function, you will probably want at least some gold and silver to barter along with your Federal Reserve Notes.

Please keep this thought in mind: the acid test

As you read what we have to say about gold and silver, even if you do not agree with what we say, please carefully consider this next point. If, for whatever reason, you decide you should get some gold and silver, please use the following as an "acid test" to make the final determination for you, if it really would be appropriate or not to buy some gold or silver:

Do not buy any gold or silver if you cannot afford to have at least a 3-month supply of everything your family needs to survive with little or no additional purchases during that period. Also make sure you can have on hand at least a 1-month supply of cash. We strongly recommend a 3-month supply of cash.

If you cannot meet both of those criteria, we believe you have absolutely no business buying and squirreling away any gold or silver.

It was one of Larry Burkett's top people, who so clearly put it another way. We saw him at a home school curriculum fair, and we spoke about gold, silver, and Y2K. He asked two very pointed questions which must be answered by each person planning to buy gold or silver:

1.) What are you planning on buying with that gold or silver?

2.) Why not just buy those things now instead?

We would add a third question and fourth question.

3.) Just who would be willing to trade for your gold and silver?

And probably most important:

4.) Who would not be willing to trade for your gold and silver?

HOW to buy gold & silver...
if you dare

As you read this whole section, please keep this next thought in mind. **We do not recommend the purchase of gold or silver**. In fact, as of the time of this writing (January 1999) we **strongly recommend against it** for many of the reasons we have already covered. With that said, we will repeat what we said earlier: if you are considering buying any gold or silver, we are trusting that any money you are looking to spend here is extra cash on hand, with all you other storage and preparation items taken care of. We also need to stress here that you should have ALL your debts paid off (with the possible exception of a mortgage on your home) before you spend one dollar on any gold or silver.

After all that, for those of you who still desire to buy some gold or silver, despite our cautions, we feel it would be wrong for us to not give you wise counsel as to how best to buy if you do buy some. With Dan's many years of personally buying and selling millions of dollars of gold and silver, he does have a significant amount of expertise that can be of great use and value to anyone wanting to buy their own gold or silver. So we felt it was very prudent we at least help you to make the best purchases possible. We will try to summarize, as best possible, some of the best advice, gathered from those 20 years of experience.

First and foremost, the lesson that too many people learn all too often the hard way (and this is an expression well know and used by many coin dealers):

There is NO Santa Claus in the coin business!

Absolutely, every time, guaranteed, if a deal looks too good to be true, is a BAD deal. In all of our experience, and in the experience of all the thousands of people we have come into contact with, this has basically ran true nearly 100% of the time. Banks don't sell $20 bills for $15, and nobody is going to sell you (just because you are just such a swell guy) gold, silver, or rare coins for less than what any coin dealer would gladly pay for it. If you are offered a deal that seems just too good, if necessary, find a coin dealer who will, for a small fee if necessary (they usually never do anything for free), check it out for you. The rare coin, gold, and silver business is full of some of the most dishonest, low–lifes this side of your local Federal Penitentiary (and that is just were some of them belong– and where some of them are today.)

We have seen people who have bought coins that were only worth less than one-tenth what they paid. We have seen coins sold as extremely rare which were actually extremely common. We have personally met too many people who have lost literally tens of thousand of dollars due to coins that were never worth more than a small fraction of what they paid. And it makes us sick.

Through our many years of experience in the coin business, we found all too many coin dealers who were on an ethics level lower than lawyers and used car dealers. And there is a reason for this. In used cars and legal affairs, there tends to be at least some degree of regulation by government officials. Yes, despite what some would believe there really is a good place for some good old-fashioned government regulation! As a result of this general lack of regulation, this business seems to be a magnet for the greedy and dishonest. True, the industry has cleaned itself up significantly from what it was like 15 years ago, but there are still too many low-lifes.

If it sounds like we are trying to strike the fear of God into you it is only because that is just what we are trying to do!

Now, we know that what we have just said will infuriate a lot of coin dealers and make us some new enemies. So we will add that there are many very honest, wonderful coin dealers out there. We knew some personally. But you will have absolutely no way to know the difference between the good and the bad. The best way to get a reference for a coin dealer is to get a reference, if possible, from someone you know and trust who has at least some idea about coins or gold and silver.

Although this next tip is not a 100%, surefire way to help detect the good dealers from the bad, it has some general accuracy to it. Long time, established dealers are far more likely to be more honest. To survive in the rare coin and bullion business requires either a lot of lying, cheating, and swindling, or a lot of happy, repeat customers. The more established dealers tend to realize the importance to surviving in business through treating people right and keeping loyal long–term customers. The more dishonest dealers tend to migrate towards the coin market when things are hot (like now) and they get out during the lean time, looking to move on to the next hot collectible, like baseball cards in the mid 90's.

We will also add that due to the independent coin grading services of the Professional Coin Grading Service (PCGS) and Numismatic Guaranty Corporation (NGC) both the degree of cheating and the quantity of cheating has gone down significantly. The grading services, while extremely far from perfect, have made it significantly easier for some who has no coin knowledge or experience to get what they are actually being represented in value, rarity, and condition.

On this subject, which we will cover in greater detail later, we give this very important warning to anyone and everyone who is thinking of buying nearly ANY coin that is not just a straightforward, bullion-value-only type of coin:

If you are buying ANY coin for any value beyond its gold or silver value, ONLY buy coins pre-graded by PCGS or NGC!

Even if you have no idea what PCGS and NGC are (which means nearly everyone) you can use the terms as sort of a mini acid test to get a feel for the honesty level of a bullion dealer. If he warns you to never buy any rare or investment types of coins (excluding bullion coins) unless they are pre-graded by PCGS or NGC, (which is what we just said) then he is very likely to be far more honest than a dealer who bad mouths the grading services and says you don't need them. The grading services are inconsistent, but they are better than nothing.

If you are unable to get a good personal recommendation for a coin dealer, the tips that we will be giving you will help to protect you. We will also add here that many gold and silver deals are pretty straightforward and the vast majority of transactions do happen with little or no problems. Most the dishonestly and fraud comes in the area of rare coin purchases where the values are based more on the rarity of the coins and less on the precious metals content. Probably the most significant issue for people buying just gold or silver will be getting the best deal for your money, and we will be covering some tips for that.

From time to time, we have been discussing the subject of collectible coins, as opposed to just gold or silver value coins. We will explain the difference between the two. To do this, we need to actually look at two categories:

- Bullion coins
- Numismatic coins

To properly understand the differences, you first need to understand the meanings of the words numismatics and bullion. Bullion is a general term that refers to precious metals. The term is used as a reference, a catchall phrase, or as substitute word for gold, silver, platinum, and/or palladium. Numismatics is basically the study of coins. It also is a general reference to collectible types of coins. More on this in a moment.

Bullion coins

There are two distinct types of bullion coins and it is important to understand the differences to be able to make the wisest purchase possible when to go to buy. Some background is needed to help you best understand the subject. By better understanding this information, you will be better able to make the wisest purchases possible for your needs. As you read the section about old U.S. silver coins, keep in mind that our personal recommendation, if you must buy silver, is to buy .999 pure one-ounce silver rounds or 100-ounce silver bars. Due to higher prices per ounce, we do not recommend the old American silver coins. In our explanation here, we will spend more time covering the silver coins, because to whole pricing structure is a little more confusing than straight, even–weight bars or rounds. More on them later.

Historically, gold and silver coins were originally produced by governments to be used primarily as basic coinage in every day banking and commerce. All but the lowest value coinage of stable governments were made out of silver or gold. Obviously, gold was use for the coins with the highest monetary value. In the greater context of history, paper is relatively new as a medium of monetary value. And even with that, except in recent years, any paper money was usually represented in value, or backed by gold and silver stockpiles held by the governments that issued the paper money.

At some point the precious metal value of the coins rose to a point that it became unfeasible for governments to manufacture their coinage in gold or silver. Eventually, the gold and silver value of the coins, became worth more than the spending value, also know as the face value. Get a good understanding of, and learn the term face value because you cannot properly, successfully buy, sell, or trade in U.S. silver coinage unless you understand it. Three quarters have a face value of seventy-five cents. The face value is the same, whether the coins are made of copper, nickel, silver, or gold. The intrinsic value, (another term you have to learn and understand) refers to the value of the metal in the coin, completely independent of any cash (monetary) value or collector value. Those same three quarters have virtual no intrinsic value if they were made out of copper-nickel any date 1965 or later. But if those same three coins were made out of the standard 90% silver alloy minted in 1964 or earlier, they will have an intrinsic value based on the weight of the silver in the coins. Obviously, due to the value of the silver in the coins, the intrinsic value far exceeds the monetary or face value the coins have as spending money.

The reference to 90% refers to the percentage of silver in the coins. They are made up of a composition of 90% silver and 10% other alloy metals. The term "junk" refers to the fact that these coins basically have no true collector's value. These are the types of coins that sometimes actually get melted down and refined to be turned into pure silver bars, jewelry, sterling tableware, or whatever else the silver is needed for. From the collector's standpoint, these types of coins are extremely common, and nothing a collector would value for anything beyond the weight of the silver content of the coin. Because these coins are sold indirectly based on weight, the general condition of them is not relevant to the value.

Although these coins are referred to as "junk silver," that does not mean they are bent, smashed, worn smooth or otherwise damaged. They will look like any other pocket change, except the color will usually be slightly whiter.

We will cover more of the valuation or pricing of these coins when we get into the specifics of buying and selling. We put most old U.S. 90% silver coins into this category of coins only worth their bullion value. In general, coin dealer correctly refers to these coins as bullion coins or as referred to above, "junk silver" or just "90%" for short.

Back to our mini History of Money Lesson 101. As we were saying, eventually, the gold and silver value of the coins became worth more than the spending value, also know as the face value. When this occurs, something happens that is referred to as "the bad money drives out the good." As people get their change, they check it to see if it is made up of silver or base metal (usually copper, nickel, or some combination thereof.) If the coin is base metal, they re-spend it. If silver, they hoard it. Eventually, the only coinage left in circulation is made of the base metal, with no true intrinsic value beyond the good faith of the general public's perception that they can spend it for merchandise at the store. All this is basically due to inflation in its earliest stages of devaluing a country's monetary standard. This is what happened to the United States monetary system. Anyone around 40 years old or older will remember the days in the late 1960's and early 1970's of being able to find silver coins in change and putting them aside to be sold at premiums over face value.

The type of bullion coins we have been referring to here were all originally made by governments for the sole purpose of being used in day-to-day commerce. This group is distinctly different then the other type of bullion coins, which we will discuss in a moment, that were made specifically to be used as an even weight of gold or silver and used for directly investing in the gold or silver itself. More on that later. Because gold and silver coins were originally designed to be used as monetary units and not specifically as measurements of weight for trade and valuation by weight, they were basically made of irregular weights of gold or silver. This gets a bit complicated, so if you do not get it the first time, re-read this next section.

Ten dimes or four quarters or two half dollars all equal one dollar. As such, a dollar's worth of each, or any combination of the group equal to a dollar (i.e. five dimes and two quarters) all contain the exact same weight of silver. **In theory, $1 of silver coins, no matter the combination, contains 0.7234 troy ounces.**

Three very important items that need to be explained here. First, this weight is a theoretical number based on the weights at the time of manufacture at the mint. Obviously, if a coin is worn nearly flat, it has less silver in it. You really need not make an issue out of the difference between the theoretical price and the actual weight. All this is factored into the buying and selling price by dealers. We only bring it up because the issue comes up from time to time and we want to let you know that unless you are stuck with a bag full of all really worn out coins, it is basically a non-issue. At the wholesale level, U.S. 90% silver coins are traded in quantities of $1,000 "face value" (the spending value of the coins) known as "bags." The normal circulation wear of coins is factored in, so instead of a $1,000 face value bag of 90% silver coins being figured at 723 troy ounces of pure silver, it is usually figured at about 715 troy ounces. A quantity of $250 face value is referred to by dealers has a "quarter bag" even if you are buying dimes or half-dollars. The term "quarter" is in reference to the face value amount compared to a full "bag" of $1,000 face value, not the denomination of the coins inside.

The second issue (this is where things get a little tricky) is the issue the amount of silver in a dollar of silver vs. the amount of silver in an actual silver dollar. You will probably want to re-read that last sentence. We said this gets tricky! So we are going to repeat this another way. In a dollar's worth of dimes, quarters, and/or half-dollars (or any combination thereof) there is 0.7234 troy ounces of silver. BUT in a silver dollar there is actually 0.7734 troy ounces of silver. Now, we *know* this is inconsistent. Don't blame us! We didn't set the standards. To learn the whys and wherefores of the matter you have to go back in time way over 100 years ago and study some of the pork barrel politics of the time. We will not go into the background any further because it is not relevant to our topic here, or relevant to your learning the ability to successfully buy, sell, or trade in silver coins. What is relevant is this next warning that we will explain later (when we cover how to figure out values): **DO NOT buy silver dollars.** Period! No exceptions! Got that? Good! Don't buy any silver dollars! (They are way overpriced.)

The third issue that you very much need to understand is the difference between troy ounces and avoirdupois ounces. To put it as simply as possible, think of troy ounces as European or metric ounces and think of avoirdupois ounces as American. Or think of it sort of like meters vs. yards in length measurements. This is very important because a troy ounce is about 10% heavier than an avoirdupois ounce. A troy ounce is equal to 31.1 grams. (Do not confuse grams with grains, which is a significantly lighter weight.) An avoirdupois ounce is equal to 28.35 grams. One avoirdupois ounce equals 0.911 troy ounces. This means there are about 10% more grams in a troy ounce than in an avoirdupois ounce. In plain English: a troy ounce of gold or silver is worth 10% more because it weighs about 10% more than an avoirdupois ounce of gold weighs.

Any dealer who tries to sell you any bullion based on avoirdupois weight is probably also trying to deceive you or take advantage of you in some way. They are usually trying to sell the avoirdupois bullion based on the troy ounce pricing. They do this to try and make an extra 10% profit based on deception. A good example of this was in some ads that used to run in newspapers and magazines offering "one-pound" silver "coins." The general idea of the ad was to promote the idea that this weighed a full pound, "16 ounces of pure silver." Sure, it was 16 *avoirdupois* ounces, but not *troy* ounces. An avoirdupois pound is actually only equal to 14.58 troy ounces, not 16 ounces. So when people tried to calculate the silver value, they figured higher than it really was. Just to confuse you even more, a troy pound is made up of only 12 troy ounces.

If any dealer offers to sell you any bullion based on avoirdupois weight, you will probably want to find a different dealer to deal with. Like our advice about silver dollars, do not buy any. When Dan was dealing in gold and silver, and he did buy any bullion that was marked with an avoirdupois weight measurement, he usually either put it with the rest of the bullion to be shipped to a refinery for melting down, or he sold it to people looking for raw gold or silver to be melted down for jewelry. But never to people looking to invest in gold or silver.

A dealer might offer you what sounds like a really good deal on some non-standard weight items, This is because those items are normally much harder to re-sell to anyone who is knowledgeable about precious metals. On the other hand, if you do end buying that type of thing, you will find that when you go to sell, you will have to take a substantial discount at the time of sale.

The good news is that this usage of avoirdupois weight measurement is rarely ever used. But the possibility of it happening does exist, so we wanted to be aware of it and understand it in case you ever run into it. All standard bullion transactions are done based on the troy ounce basis of weight. Virtually all bullion coins and bars are made using the troy ounce basis for weight. So you should not have to worry. For the usage in the rest of this article, **when we refer to ounces, we will be referring to troy ounces only.**

The second type of bullion coins are coins that are manufactured for the distinct purpose of being bought and sold for their gold or silver value exclusively. No collector or rarity value intended. These can be made either by businesses or national governments, like our own federal government. (All standard government issued gold and silver coins are based on the troy ounce basis of weight measurement.) These types of bullion coins are almost always made in even one troy ounce weights, or standard fractions of a troy ounce: one-half, one-quarter, and one-tenth troy ounces. Some of these coins will be marked ".999 fine gold." Some may be marked slightly different.

Purity

Before we continue on this subject, we need to cover the subject of "purity" as it relates to gold and silver. The purity factor is something you do want to understand when it comes to buying any gold. When so-called "pure gold" is mined in the earth, or "pure gold" nuggets as panned by miners in a stream, they are actually only about 75-95% pure with various impurities mixed in. This gold is sent off to the refinery, where the impurities as refined out. Through the refining process, the gold becomes what is known as "pure." What is meant by "pure" is that it is usually 99.9% or 99.99% pure. For accuracy sake, from a technical standpoint, it is almost never referred to as 100% pure. Even gold from the most reputable sources will not be marked as 100%, due to minuscule amounts (one part pure ten thousand) of impurities that are practically impossible to remove absolutely, completely. Something that is worth keeping in mind this that most well recognized forms of investment gold or silver will be marked with something like .999 fine gold or .9999 fine gold, but almost never 100% or just the term "pure gold" or 24K. The same is true for silver. (Obviously, silver never would be referred to as 24K, only gold would be marked that way.) If it does say 100%, that is actually something to be suspicious of.

The term 24K means that of 24 "parts" of the gold, all 24 "parts" are "pure." (The K stands for the word "Karat.") In contrast, 10K gold is 10/24th "pure" gold and 14/24th non-gold alloy (10 + 14 = 24). To figure out the percentage of purity, take 10 divided by 24, which equals 41.66% pure. Another example is 18K gold that is 18/24 pure gold plus 6/24 non-gold alloy for a total of 24K. 18 divided by 24 is 75% pure. (OK, so this doesn't have anything to do with

Y2K! But is does have a lot to do with gold, so we threw it in, at no extra charge, for those of you who were interested!)

The most common purities for silver are .999 pure, 90% silver coins, and .925 (92.5%) pure for sterling. We will say this about sterling. Never buy any sterling silver. When you go to sell it, you will find that it is very hard to sell for a good price, and all the buyers want to pay steep discounts for it.

Bullion coins that are manufactured by businesses, as opposed to those made by governments, are generally referred to by the phrase of "privately minted." The term "mint" refers to two things. In this case mint refers to the manufacturing process of making coins and the name of the building where coins are made. In the other context, "mint" refers to a condition or level of preservation. A mint condition coin refers to a coin that is basically in new, unused condition. More about that in the numismatic section.

Here is a good tip to follow when buying bullion coins; only buy coins that were minted by official governments like United States Gold Eagles, Canadian Gold Maple Leafs, South African Krugerrands and such. Avoid buying gold coin minted by non-governments. (The only exception to this is gold from Johnson Matthey refineries or the Credit Swiss bank and even these two are less desirable than gold from the governments we just listed.) A follow-up tip is to avoid buying any gold or silver bullion style of coins that have any type of premium to the bullion value which is in any way tied to rarity, being a limited edition of any kind, or tied to any kind of collectible premium values. All you want to be paying for is the basic bullion value plus a fair handling charge mark–up. We will get into what constitutes a "fair" handling charge in another section. A very common practice of both private and government mints is to produce quasi-type bullion coins in limited editions, low mintages, with "collectible" themes, or with some other gimmick to try get some type of premium price over and beyond what similar straight forward bullion coins trade for. Avoid all such coins like the plague. We will tell you that from all our experience, more people who buy such types of "collector" or limited edition "so-called-rarities" end up selling them for little or no premium beyond the straight bullion value. These are the types of coins that some of the less than honest coin dealers try to sell because they have significantly larger profit margins on them. This

233

translates into more of your money into his pockets and less of his gold and silver into your pockets.

Time to buy and sell

At this point, before we continue, we would like you to pause for a moment and think about the term "buy/sell spreads." We previously covered what "buy/sell spreads" are in Chapter 10 in the liquidation section. If you do not fully and completely understand what "buy/sell spreads" are *and* how they work, please go back and re-read that section. It is extremely important to know and understand for you to get the best values when you go to buy and sell coins or gold and silver of any kind. Understanding this can also help you for buying and selling many other things.

Buying precious metals is generally a lot more involved than going to the grocery store, getting two pounds of bananas for 39¢ a pound and paying the cashier 78¢ for the bunch and walking out. There are lots more variables to factor in. For starters, you are not going to sell your bananas; you are going to eat them. Now that statement may seem so obvious that it is foolish, but the selling factor makes a major difference in the purchase. With bananas, you just figure out who in town has the best price, you buy, take home, and eat. You do not have to worry about if they are 90% pure or .9999 fine. Fractional weights vs. even ounce weights are not an issue. Where they were made (grown) does not matter. And whom you will sell them to is a non-issue. But these issues are very relevant when you buy gold or silver.

For our examples, we will be using nice round numbers as much as possible, to help keep the focus on the concepts and terms we are trying to teach. We will be using some general numbers for values, but the weights in the example will be accurate. The prices and values listed reflect current market pricings as of January 1999. We will be referring to a term called "premium." For our example, the premiums will be generalized, because "premium" values fluctuate with supply and demand at any given time in the market.

Premium

So what does the term "premium" mean? No, it is not a free coin found in your cereal box. A "premium" is the term used in the coin gold and silver industry to describe the mark-up in price over and beyond the straight value of the pure precious metal. If the price of an ounce of gold is at $290 in the trading pits of the commodities market, but a particular one-ounce gold coin sells for $310, it has a premium of 6.7%. Do NOT get the term premium confused or mixed up with the term "mark-up" or "profit." If you buy this one-ounce gold coin for $310, it does not mean that the dealer selling it is making a $20 profit ($310 minus $290.) He may have had to pay $302 to buy the coin, thus making only $8 on the transaction. On the other hand, if you needed to sell that same coin, he may only pay you $293. We will explain how this works as we cover buy/sell spreads.

Profit Margins

For many transactions with most reputable dealers, the profit margins run very small, from 1-2% for larger deals, to 10-15% at a smaller transaction. Profits of much over 15% are only seen on smaller transactions, such as those under $100. We will cover "mark-ups" here as we discuss "buy/sell spreads" in more detail. However the exact percentages will vary on a dealer-by-dealer basis.

Spot Price

Another term to learn is the "spot price." The spot price refers to the trading price per ounce of gold or silver on the New York, Chicago, Zurich, or Hong Kong commodity exchanges. Of course they are not trading the actual bars of gold or silver, but the paper contracts that represent the value of a specific weight of bars of gold or silver. You will hear the term "spot price" referred to quite often if you buy or sell any bullion.

Buy/Sell Spreads

To truly make the best purchase, to get the best value, you need to understand both buy/sell spreads *and* premiums. Just knowing something has a buy price of $293 and a sell price of $310 does not have anything to do with the premium you are paying. Furthermore, you have to understand there are actually two buy/sell spreads: yours and his. This gets a bit complicated here, with all the numbers, so re-read the next few paragraphs if you need to fully understand.

Using the $310 gold coin example from above, we will explain all the numbers and relationships.

First, premiums. Any purchase of gold or silver needs to be done with the premium understood and factored in. You need to understand and know how much of the money you spend represents the actual bullion value or "melt" value of the coins involved (weight times the price of gold or silver. In our example, the gold coin has a "melt" value of $290.) You also need to know how much of your money is going to pay for various premiums. In our example there was a $20 premium. We are going to ignore the issue of sales tax for this example. We will deal with that later. There is no way to say how much of a premium is fair or reasonable unless you understand the buy/sell spread. A 15% premium might be a good deal, if dealers are paying 10% over melt, as sometimes happens with 90% junk silver coins. At other times, junk silver actually sells for less than the melt price! It all depends on *supply and demand.*

Back to the example. Once you buy your gold coin for $310, you ask the dealer how much he would buy it back for, if you need to sell it, assuming the spot price is the same. He tells you his current buy price is $293. That is $17 dollars less than you paid. However this does not mean that he just made a $17 profit on the one he just sold you. Keep in mind that just because his buy price is at $293 does not mean he can buy all he needs at that price. If you buy 20 ounces, he will need to find a seller if he wants to re-stock his inventory. Most dealers will want to re-stock immediately, for reasons we will not go into here. This means that he has to find a seller on the wholesale market (unless, of course he just happens to have a different customer walk in about the same time wanting to sell 20 ounces of gold, which almost never happens.) He gets on his electronic dealer network, or calls his wholesaler gold and silver broker. He finds out that for dealer to dealer trades, the buyers are paying $299 and sellers are asking $302. Thus, the wholesale buy sell spread is what is spoken between dealers as "$299 *at* $302" which in plain English says, they will pay him $299 for his or sell to him *at* $302. When he re-stocks his inventory at $302, he only makes a $8 profit per coin. Not a huge mark-up!

That was the simple example. Now we make it a little more complicated. The goal of our next example is to show you why it is better to buy gold in one-ounce coins instead of smaller, fractional ounce coins. To help make this shorter, we will start by summarizing our fractional gold example by telling you this next item. The retail buy/sell spreads on fractional gold coins is wider than the buy/sell spreads on same style ounce gold coins. This means that if you decide to buy gold coins, you will get less gold for your money if you buy fractional weight gold coins than if you buy one-ounce gold coins. Or put another way, figuratively speaking, fractional means more profits for the dealers pockets, and less gold for your own pockets.

In addition to the larger commissions for both buying and selling fractional weight gold coins, there is another factor why fractional weights are not as good of an investment. The more money invested in premiums means the less gold (or silver) to go up in value. Using our previous set of numbers, if you buy ten ounces of gold for $3100, you have $2900 actual gold value. If gold were to double in price, you would have $5800 worth of gold. If you bought the same dollar amount of gold in one-tenth ounce fractions, it would work out like the following. Tenth ounce gold has a premium of about 15%. That puts the price of one-tenth ounce gold pieces at paying $333.50 per ounce instead of paying $310 per ounce for once ounce piece. With the same $3100, at $290 per ounce, you can buy 9.2 ounces of one tenth ounce gold coins for a total cost of $3068.20 with an actual gold content value of $2668. In this case, if gold prices double, that $2668 turns into $5336. Compare the numbers. $2900 worth of gold melt value if you buy one ounce coins, compared to $2668 if you buy the tenth ounce gold coins. If gold prices did double, $5800 vs. $5336. The choice should be clear.

Of course, there is the argument that fractional gold would be better for barter. We disagree with that argument. For starters, think back to our arguments as to why we are very unlikely to go back to a non-dollar based barter system. But even if we did go to a barter system, one-tenth ounce gold pieces would still be too hard for many, if not most, barter trades. That is where one ounce silver coins (or "rounds" as the dealers refer to them) and 90% junk silver coins come into play. Silver is the best for bartering. Gold is best as a store of wealth, value, assets, and such. Gold is best for being able to store a greater amount of value in a smaller location, or greater transportability of a larger value in a more convenient, discrete, smaller size. The objective should be to come out with more gold as the sacrifice of smaller unit sizes.

A really good salesman will try to tell you the example we gave is not exactly accurate, because when you go to sell or trade, you will get some of that premium back on the fractionals. But when you go to sell or trade the one-ounce coins, you will get less of a premium.

We will tell you what Dan's experience was in reality, having actually bought and sold millions of dollars all forms of gold and silver when the market was up and when the market was down. When the market is down, sometimes the fractionals can be sold for premium over the spot value, but the premiums are not all that much. When the market is soaring high, almost nothing trades at much of a premium when it is time to sell, because high prices tend to "absorb" any premiums, and buyers tend to be buying closer to the melt down values, one ounce or tenth ounce. IF things do come down to bartering, people will not give you an extra premium value on you fractionals, just because they were worth a premium. The will just figure them at whatever the current gold value is multiplied time the weight. Or put another way, everything will likely be figured at melt, without a discount. The exception to all this will be for any non–standard, odd, or uneven weighted coins, or for very uncommon gold or silver coins. They will be more likely to trade at actual discounts to the melt value.

If you do buy any gold coins, do not carry them around in your pockets or store them loosely. Gold is relatively soft for a metal, and a gold coin will get scratched up in less than a day of being carried around loosely in a pocket or purse. If you try to sell a scratched up gold coin to a dealer, you will be discounted up to about 5% off what you would have got otherwise. Most dealers will put the coins in a good protective container. If you do not have an appropriate coin holder to transport a coin in (i.e. you bought ten in a coin tube, but you only want to bring one in to sell and you do not want to bring in the nine others with you,) just wrap it carefully in tissue. Use the soft kind of tissue, like you would use to wipe a tear with, not the stiff, scratchy kind like is used to wrap presents with.

When you are buying gold coins, ALWAYS examine each gold coin for any serious blemishes (a few small nicks or scuffs are OK, if not too serious or distracting) before they are put into the container. If you are handed a container with the coins already in it, ask the dealer to allow you to quickly examine and count the coins, *in the dealer's presence*, before you actually accept them. Be polite but firm. If you use good manners, the dealer will respect you for it, unless he is just a jerk. (Sorry to use such a harsh term. Most dealers are very polite, even if they are dishonest. But let's face it, there are some real jerks out there in the coin business.) Usually, there is never a problem with the count or with any seriously damaged coins, but if there is, you will want to deal with it before you actually "take delivery."

This same caution about examining and counting the coins can be said for buying silver as well. However, with silver, a lot more scuffs and scratches are acceptable. Just be checking for serious damage. Because silver oxidizes, a little tarnish is okay, but do not buy anything, other than 90% junk coins, if there is more than about 25% coverage of tarnish. This is a general rule, not a hard, take-it-or-leave-it type of rule. But if you do have silver bars or coins (again, this is for other than 90% silver U.S. coins) that are heavily tarnished, you will find that in general when you go to sell, the buyer will want to pay to slightly less than he would for the same thing if it is untarnished. With 90% junk silver coins, tarnish does not matter too much unless there are a lot of very heavily tarnished dark coins in the group.

When you are buying 90% junk silver coins, there are a couple things to keep in mind. When buying junk silver, most dealers will count it out with an automatic coin counter, put it in a cloth bank bag (the type they carry around large amounts of money in the movies), put a label on it indicating the face value dollar amount (not how much you are spending), and then seal it with a tamper proof seal. This is done so that when (if) you go back into that same dealer to sell, he will know exactly how many dollars of face value of silver coins are in that bag, without needing to count it first. When buying back junk silver they sold, with their own seal on the bag, sometime dealers will pay you without the need to re-counting or study the contents. There is another little benefit to leaving the bag sealed.

Going back to the actual purchase of your junk silver, there is one request you should make after agreeing on the price and agreeing to buy. Politely ask the dealer to let you watch the counting and the sealing of the bag. Explain to him that you would like to watch the actual counting and sealing, so that when you can say to whomever you sell it to that you actually saw the counting and the sealing of the bag. And so that you can say that you know the coins really are 90% silver, not current coin, and they the are not a brunch of worn out and damaged coins. If you ask politely and tactfully, most dealers will comply with your request. If they will not comply, consider finding another dealer who will. Asking to be able to verify how many coins you are buying is not an unreasonable request.

If you have the option of your choice of denomination you want to buy (dimes, quarters, or half-dollars) we suggest quarters or half-dollars. If you ever get in a situation of needing to count out a larger amount of these coins buy hand, you will be more than glad you did not select dimes. We know this from personal experience! Some people recommend getting dimes because the are smaller value amounts for bartering. If you ever did get in that situation, you should be able to trade some of your larger denomination coins for and equal value of smaller denomination. Personally, we would not worry about that aspect as we have already covered it.

Silver options

Before we can get into the details about the various silver options, we need to give you one piece of warning about buying silver. Like our advice to you feel you just must buy silver, do not buy 90% junk silver coins. We will be covering the reasons why later in this section. Further more, if you already own and 90% junk silver or silver dollars, if the premiums are still high when you are reading this, sell them! If you still want to be holding silver for investment or bartering, buy 1-ounce silver rounds or 100-ounce silver bars.

We have already explained some about 90% junk silver coins and one ounce silver rounds. But there are more details that need to be understood to be able to make the wisest possible silver purchases. The most common forms of silver are 90% junk silver coins, one ounce silver rounds, and silver bars of 1–ounce, 10–ounces, and 100–ounce weights. The most relevant factor to consider here is the balance between getting the most silver for the money and still have good liquidity at the same time.

To best understand the factor of getting the most silver for the money, some background information is needed. We have already covered the issue of just how much silver is in a dollar face value of U.S. 90% silver coins. But by itself, this weight information does not reflect value. If you take the weight and multiply it times the spot price of silver, you come up with what is called the "melt value" of junk silver coins. Knowing the melt value will give you somewhat of an indication of the market value of junk silver coins. But as with the previous examples of the gold coins, knowing the melt value of these coins does not indicate what the actual buy/sell spread is in the real world. To determine actual buy/sell market value of junk silver coins you need some sort of access to that trading market.

With most gold coins you can usually know the buy/sell spread through knowing the spot price and multiplying by the appropriate percentage factor. Example: one ounce gold eagles generally have a dealer buy price of anywhere from the spot price to as much as three percent over spot. Sell prices will run between 4-7% over the spot price of gold. The ranges are determined by a combination of the quantity of ounces traded (the more coins traded, the better the price) and the demand within each

With silver, there is a lot more variables. Part of that is due to the greater variety of options. But in the case of silver, demand has a tendency to produce much greater swings in the percentage of premiums charged. We have seen silver bars and one ounce rounds have a range in the wholesale premiums varying from as much as one dollar over the spot price all the way down to actually trading at slightly *below* the melt down value! This same scenario also occurs with junk silver coins.

It is very hard to explain the complicated reasons for these wild fluctuations in the silver price premiums, relative to the spot market price. We will try to do it as simply as possible. But keep in mind that what is less important than why this all happens, what is more important is what to do about it, for the greatest profitability.

As we previously mentioned, the spot prices are based on the commodity trading prices worldwide. This is on an extremely macro-economic basis, with wide global influence based on the combined supply and demand of the entire world on any given day. The spot price is arrived as the basis of literally hundreds of millions, or even billions of dollars worth of actual trading for that day.

Of course, the spot price has a major influence on the general pricing of gold and silver bullion. But there is an additional factor that plays into effect for junk silver coins. That additional factor is in the form of the supply/demand as it specifically applies to junk silver coins primarily in the United States alone. You see, when they trade silver contracts globally, they can care less about U.S. 90% silver coins.

Dan first learned about this whole process from a good customer at Dan's first coin dealer job. There was a very good customer who came in one day and said he wanted to trade all his junk silver coins in for 100-ounce silver bars. Dan explained to him that he would lose some money, due to the buy/sell spreads. The customer then explained that although he has losing a little bit of money on the buy/sell spread, because of the wide differences in the premiums relative to the spot price, he would actually be walking out the door with significantly more silver in his possession. At that moment, Dan's entire perspective of investing in gold and silver changed. This guy was right! By playing the differences in the premiums, over the long run, a person could actually accumulate more silver without actually having to spend any more money! Dan worked with many customers since then, whom he taught this same lesson too.

The relevance to you about what happened way back then in that coin shop is that you do not want to be the one buying silver (or gold) on the wrong side of that extreme pricing variance. Do not buy the bullion with the high premiums. If you feel the need to buy bullion, buy it in a very common form that has both the lowest premiums yet is still well recognized as a standard bullion industry form. Example: buy bars or rounds in even weights, not a sterling silver tableware set weighing 136.72 ounces, even if it is at a big discount. Using the same logic and advice, we recommend you do not buy junk silver coins. We will go through all the math so you can see how the numbers prove our recommendation.

As of this printing, the current buy/sell spread on $1,000.00 face value bags of 90% silver coins is buy at $4,800.00 and sell at $5,000.00. In coin dealer terms, this is called 4.8 times face at 5.0 times, also called 4.8X at 5X (the "X" as a shorthand reference to the word "times"). What this word "times" means is that for every dollar (or any portion thereof) face value of silver, they are paying $4.80 cash value, per every dollar of silver coins. Or as the term implies, 4.8 times $1.00.

This "times" word and "x" term are important to understand if you want to successfully buy or sell any junk silver. There are a number of reasons why this is important. For starters, if you do not properly understand the terms you can not effectively communicate. It also labels you as a novice which means you are much less likely to be offered as good of a deal. Even more importantly is if you know the terms and you can effectively communicate in the terms of the dealers, it is significantly less likely the dealers will try to cheat you or take advantage of you. A dishonest person usually only tries to cheat people who do not know they are being cheated. One of the main reasons for this is no one would try to try cheat someone if the victim knows they are being cheated.

Back to the example. A dealer who has to pay $5,000.00 for a bag will ask about $5,200.00 for it. On the other hand, their buy price on the same bag would be about $4,600.00. This bag contacts, at best 723 ounces of silver. At $5.00 per ounce that puts the melt, cash value of that $5,200.00 dollar purchase at only $3,615.00 ($5.00 X 723 ounces.) If the spot price of silver doubles (which we do not believe will happen), the melt value goes up to $7,230.00.

The dirty little secrets of silver coins & 1-ounce rounds that your bullion dealer doesn't want you to know!

When the doomsday crowd is advising you to stock up on junk silver coins, they fail to let you know of the steep premiums you pay. They fail to tell you that when either the prices go up and you want to sell, or all the Y2K buying frenzy is over, and everyone else wants to sell. Those big premiums on the buy side will turn into steep discounts on the sell side. They also fail to tell you of the advantages of 1-ounce silver rounds! At the time of this printing, with silver at about $5.00 per ounce, 1-ounce rounds sell retail for about $5.75 each. If you buy 1-ounce silver rounds with that same money as above, you will get 904 ounces of silver worth a melt value of $4,520.00. If the price of silver doubles, that becomes $9,040.00 melt value. $9,040.00 in 1-ounce rounds compared to $7,230.00 melt value of 90% junk silver coins. Which sounds better to you? Keep in mind that 1-ounce silver rounds are also excellent barter items. Although not quite as small, incrementally, they still offer all the same advantages. In fact some people actually prefer 1-ounce silver rounds to junk silver coins! Because they are able to more easier figure the exact silver weight involved in the trade, they are better able to determine values and therefore if it is a good or bad deal.

(In the words of those wonderful TV info-mercials...) "But wait, there's more!"

Based on twenty years of experience either in the gold and silver business or following it, we can tell you that the higher the price of silver gets, the smaller the premiums for junk silver. This factor continues to operate to the point that historically, whenever silver prices got really high, junk coins actually traded at *discounts* to the melt value! So if you pay a 33% *premium* on 90% silver now, if the price did actually double (again, highly unlikely) you could be finding yourself actually selling at a 10-20% *discount off* the melt value!

On the other side of the coin (pun intended), 1-ounce silver rounds, as well as 100-ounce bars, tend to trade at prices consistently closer to the actual spot prices. Remember, when the prices go up sharply, it is on the basis of worldwide demand. This worldwide demand is for the contracts first (as what is traded in the commodities pits, where the spot prices are fixed) and then for the pure thing, second. Internationally, although known and recognized for what they are worth, there is little demand for U.S. junk silver coins.

There are two types of one-ounce rounds: Americans Silver Eagles and what is called "generic." American Silver Eagles are made by the U.S. Mint. The government only sells them wholesale to a limited number of major bullion dealers. They sell in lots of 1,000 ounces for spot plus one dollar per ounce per coin. These major bullion dealers turn around and resell these Eagles to all the other dealers in the country for about $1.10 to $1.25 per ounce over spot. These dealers are the ones who in turn sell to the public for about $1.50 to $2.00 over the spot price.

Of course there is also what is referred to as the "aftermarket." When the general public decides that it is time to sell, they walk into a dealer somewhere and sell. If the dealer can not sell them in his own store, he either sells them to his wholesale bullion dealer or he finds another dealer on the electronic dealer network.

The market for gold Eagles operates the same way.

"Generic" is a term for silver that is not "minted" by a specific, well-recognized manufacturer of bullion. Outside of our own government's mint, the two most recognized silver processors are Johnson Matthey and Engelhard. Silver from these two refineries will trade for slight premiums to similar silver from other manufactures. Most generic one-ounce silver rounds sell wholesale for about 35¢ to 50¢ over the spot price.

If you do the math, you will quickly figure out that you get a lot more silver for your money if you buy generic silver instead of silver Eagles. Sure, you would get a little more for the Eagles when it is time to sell, but that is only because you paid more. The problem is that if silver does go up in value, only the silver appreciates in value, the premium does not. So every dollar spent on premiums is a dollar spent that will not go up in value. Keep this same argument in mind as you read the next section on silver dollars.

Bottom line: if you are looking for a good store of wealth in the form of silver, 1-ounce rounds and 100-ounce bars offer you the most investment bang for the buck. You pay a very steep premium for the smaller individual trading units that 90% silver coins offer.

If you already own a quantity of 90% silver coins, we would suggest that you trade them in for one ounce rounds. Call around to different dealers a find the best deal you can get for getting the most rounds you can trade for your 90%.

Silver dollars

Earlier in the chapter, we told you to never buy silver dollars. If you follow the math, you will see why. Silver dollars have a very high collector's value relative to their silver content value. A silver dollar contains 0.7734 ounces of silver for a melt value of about $3.90. Due to the collectible nature of silver dollars, they currently sell for nearly triple the melt value at about $10 a piece. If the price of silver does double, you can be sure that silver dollars will not double in price. If they even went up by 50%, we would be surprised they went up that much. They are a nice collectible that is interesting to own, but a lousy investment.

This same argument applies to any kind of old U.S. $5, $10, or $20 gold pieces. Don't buy them! They sell for incredible premiums over the meltdown value. A $20 gold piece has just under one-ounce of gold in it, yet they sell for over $500. That represents a premium of more than 80% over the gold value. Since most of you do not know much about old gold coins, that may not be such an unreasonable premium for such an old, rare coin. Well, the fact of the matter is that these old, used gold coins of this type are actually very common. And except for all the hysteria caused by Y2K, they normally only sell for slight premiums over the gold value. This also has been historically true for most $5 and $10 gold pieces as well. Until now. With all the abnormal Y2K hysteria and hype, demand has been extremely unusually high due to the snake oil sales promotions. (Refer back to our "Beware of the snake oil salesman; he has a mean bite!" article.) As we said, don't buy these unless you don't mind only getting half your money back.

If after all this, you are still not convinced, and you still plan on buying old U.S. gold pieces, we will give you this advice. This advice will best protect yourself and your investment from the least financial downside risk. The coins with probably the smallest buy/sell spread, the best liquidity, and (we think) the least downside price risk are MS 63 $20 St. Gaudens or the older $20 Liberty gold coins graded MS 62. Again, our previous rare coin advice applies here: buy only coins graded by PCGS or NGC. If you need more information about buying these types of coins, consult a coin dealer.

Negotiation tips

Just a few pages ago, we referred to the importance of understanding the terms that bullion dealers use. As we mentioned, by understanding these terms, you are less likely to be taken advantage of. There is another side to this coin. Not only are you less likely to be taken advantage of, but you are also likely to get a better deal if you know just what to say. One of the things we have learned in the coin business is this strange little quirk that seems to be fairly consistent among most coin dealers and bullion dealer: the more knowledgeable you sound, the more competitive dealers are on pricing. Dealers pretty much understand that the more knowledgeable a customer sounds, the more likely they are to shop around for price, and the more likely they are to be aware of where the real wholesale market buy/sell spreads are. So the unconscious assumption is that if the dealer wants to get the deal, he better offer better pricing than he would give to someone who is completely "retail." While this is not true 100% of the time, it is true a great deal of the time. Where this is less likely to be true is at some of the much larger dealers who just set their prices and are less likely to deal. It is more "just take it or leave it" with them.

Other good terms to use in referring to silver coins is "junk silver" or just "90%." When asking for prices on 1-ounce silver rounds, just use the phrase "1-ounce rounds." The dealer will know what you are referring to. Dealers always use the shortest, abbreviated terms when talking amongst themselves. As you are calling around to dealers, listen to the terms they are using and right them down. Use some of those terms. Just be careful not to use any terms that you are not positive what they mean!

When you are ready to buy, and you want to shop around for the best pricing, it is best to wait until after the bullion market in New York closes for the day, about 3:30 eastern time. This way, everyone you call will be using the same spot market price basis to compare pricing.

Another little tip is how you ask for the price. If you *first* ask what they are paying, they might give you a lower selling price, in order to sound more price competitive, than if you just called in asking right off what they are selling for. If you are looking to buy some junk silver, first ask, "what is your buy price on a bag of 90%? (Wait for answer) Well then, what are you selling at?"

These same approaches can be used for trying to get a better deal on any bullion you are looking to buy or sell.

Making the purchase

Once you have finally decided what to buy and you have agreed with the dealer on the price, it is time to pay up. Whereas almost anywhere you shop, businesses will take a check. However, in the gold and silver business, unless you already know the dealer and you have established credit with them, they will rarely ever take check for anything except smaller amounts. The amount over which dealers will not take a check varies from dealer to dealer. But the general practice is universal. And dealers never take credit cards, unless you are willing to pay an extra 3-4% handling charge to cover the additional bank fees the credit card companies charge.

If you are going to buy in person at a local dealer, it is best to call in advance to see if he prefers cash or cashiers check. Some dealers do not like to handle much cash because of government cash reporting regulations. If you are so blessed and have so much wealth that you want to spend $10,000 or more on gold or silver, do not use cash. Most dealers will absolutely not take that much.

As far as trying to ask for a cash discount, if you pay in cash, forget it. You will either insult the dealer or annoy him.

If you are buying though the mail, once you find who you want to deal with, ask how he prefers payment and see if you need to set up an account before he will take you order. Just like a stock broker requires an account be set-up first, before they can accept phone trades, bullion dealers also need accounts set up in advance before they can take bullion phone orders from complete strangers.

That is all we have in this Handbook on the subject of buying gold or silver bullion. Your local dealer is likely to have books on the subject if you want more information.

Numismatic "investment" coins

We have only three words to say in reference to buying any coin worth any value beyond it's basic melt value plus a reasonable handling premium (such as with 90% junk silver). Just three simple, easy to follow words that are guaranteed to save you grief, trouble, and money. As far as anything collectible or of "investment quality," our three secret words are:

Just say no.

That's it.

But we know that this simple advice will not be sufficient for at some readers out there. So for them, we have this next section:

Is numismatics a good investment?

For those last two hold–outs reading this, we will start with this question:

If professional coin dealers with years of experience have a hard time making profits on investing in coins, what makes you think you can do it when professionals have a hard time doing it?

The fact of the matter is that as a true investment, in general, coins stink. And any truly honest coin dealer will admit this fact. Oh sure, there are some coins that have done very well. In fact there are some coins that have gone up tens of thousands of percent since just the 1970's. Ya, sure. And if we bought Microsoft stock when it first came out, we would not have to work for a living. Or if we bought stock in AOL the first time we saw one of those AOL disks with the free 10-hour trial period, we would have made more huge profits. Every investment group has those few items that have absolutely stellar performance. And then there are the things that everyone else buys. Any honest coin dealer will tell you that as a general rule, the coin market has for the most part gone down or been flat ever since the summer of 1989. Until Y2K, of course. Because people will not have power or food or water, they will need rare investment coins! So rare coins are sure to be a great investment in the year 2000, as our economy collapses around us! (Okay, we will remove our tongue from our cheek now.)

So if coins are such a lousy investment, how do we explain all those rich coin dealers? With one word: "Mark-up" Money and wealth is not made in the coin business though buying a coin for $500 and waiting until the market goes up 50% to sell it. No. They buy the coin for $500, mark-up the price to $600, sell it as fast as they can, taking a quick profit and moving on to the next coin. Now the guy who just bought the coin will likely never make a profit on it, but the coin dealer sure did. See, investing in coins was very profitable for him! Now if the poor guy who bought the coin tries to sell in, unless the market moves up, he will only get about $350 to $450 depending on if the can find a dealer actually looking for this type of coin.

Another dirty little secret about investing in rare coins is that most "investors" lose money. Usually lots of money. With rare exceptions, most people we have known have either lost money or basically broke even. The very few who actually made good profit did it through lots of studying, following the rare coin market very closely, and lots of experience the rare coin school of hard knocks (they lost lots of money at first!)

Still want to invest in rare coins?
Okay, well this is where you start.

The next few pages we are re-printing from some information that Dan used to distribute to all his own coin customers and anyone else he could get to read it. It is still good advice for today.

How to not get ripped off when buying coins

The Problem

The greatest concern for any investor or collector is not the most obvious: "Will this coin appreciate in value?" The greatest concern is first and most importantly: "Is this coin correctly represented, both as to the condition and the price?" Compared to the few people who ask me if I think a specific type of coin will go up in value, many more people ask me: "How do you grade this coin? If I were to try to sell this, how much would I be able to get for it?" Most people are already aware of these two basic facts: 1) Rare coins have historically been good investments. But 2) Quite often the value of a specific coin offered for sale is misrepresented; either by grade or by price. Hence, the biggest challenge for most collects and investors is "If I am not a coin dealer, who travels to lots of coin conventions and has years of experience, how can I find investment quality coins that are both accurately graded and fairly priced?" When you buy coins, you need to have an assurance that you are not getting ripped off and assurance that you can sell your coins as the same grade you bought them. Buying only independently certified coins can give you that assurance.

Very few investors or even collectors will be able to grade as well as the people they buy coins from. I've seen lots of people try and the vast majority end up failing. I have had countless people show me what they believed was a "great deal" they bought from a dealer or a so-called "private party" (who was also, in all likelihood, a dealer or very sharp, experienced collector.) The coin almost never turns out to be as great as they originally thought it was or were lead to believe it was. In nearly every case the coin had subtle but significant defects such as light cleaning or hairline scratches. It takes a great deal of learning and experience to accurately grade coins and detect these kinds of defects, and very few dealers will take the time to teach it.

Never forget this next line, because it is the most important one in this chapter: "There is no Santa Claus in numismatics." I am not saying you can't find uncertified coins that are graded and priced accurately!! But, if deal looks too good to be true, then that's just what it is: too good to be true! If any coin is offered to you at a real "bargain" price, be very suspicious of it. There is no such thing as accurately graded coins that are offered at "below dealer cost" or "below wholesale prices" and don't ever forget it. If the coin were really that grade, they could very easily sell it for more to any dealer with a lot less trouble.

The Solution

The first step in purchasing accurately graded and priced investment quality coins on your own is to acquire coins that have been independently certified by one of the following services: Professional Coin Grading Service (PCGS– currently the most common and popular grading service.) or Numismatic Guaranty Corporation (NGC– these coins tend to trade at price levels similar to PCGS for common coins and are discounted up to 30% on more scarce coins.)

These services give you a third party opinion of the grade by someone who is neither buying or selling the coin. They will lower the assigned grade for the coin if it has a value-reducing defect. Light cleaning, rubbing or hairlines may not be obvious to you, but are very noticeable to an expert's well-trained eye. When you buy a certified coin, you have the assurance that "what you see is what you get." Most reputable dealers will buy or sell the coin at the same grade as assigned by these grading services.

In addition to the grade/price reassurance that a PCGS or NGC graded coin gives you, certified coins have two other benefits when you go to resell them: increased liquidity and a premium in the price you receive.

PCGS, NGC vs. the other grading services...

I believe the two best grading services are PCGS and NGC because these reason:

1) These graded coins trade at much closer buy/sell margins compared to other services.

2) Currently PCGS coins are the most liquid for resale because there are so many dealers who offer to buy many types and grades of PCGS coins on a sight–unseen basis. This means that there are hundreds of dealers who will quote you a price they will pay you for a coin in a specific grade if the coin is in the PCGS holder– without even seeing the coin first.

3) It sometimes seems that you have to pay more, but certified coins are usually worth the premium, and generally much better quality than those offered at similar prices, but are not certified. Just think of that extra premium as being like insurance– insurance that the coin is accurately graded.

4) Overall, you will find that most larger, reputable dealers are much more interested in buying coins graded by these services, and pay higher prices, than for the same coin either uncertified or certified by some other grading service.

Why some coin dealers don't like NGC or PCGS

Question: If these grading services are so great, why do some coin dealers dislike them so much?

Answer: Why do crooks dislike police?

While this analogy does not hold true for all coin dealers who don't like these services, it does hold true for the majority and especially of those who are the most vocal about it. Let's take a fictional "Honest John's Coin Shop" as an example and look at some of his claims. "Those grading services overgrade coins." Or, "I will sell you coins that are nicer for less money." If he really believed that, he would put his money where his mouth is. He would send in all his "undergraded" coins to get them all graded at higher grades by these so–called "overgrading services" so that he could sell them to practically any dealer in the country for double or triple the price he's asking of you. But noooo, "Honest John" is such a nice guy that he is going to sell you his non–certified coins and pass on all his great bargains to you! The fact is that if all of "Honest John's" coins were sent in to NGC or PCGS, about half the coins would be rejected, ungraded due to cleaning, altered surfaces, or some other value reducing defect. The other half would probably come back an average of one to two grades lower than what he would sell them to you as...for bargain prices of course!

On the other hand...If you were selling your coins, Honest John will tell you not to waste your time or money on grading services. He'll say "I will buy them at higher grades than what they will grade them." Or "My friend sent in ten MS 65 coins and they all came back MS 63 or lower. Don't waste your money." Of course he doesn't want you to send in your coins if you are selling, because then he can't buy them at lower grades than what they really are.

No, PCGS and NGC are not perfect.

I have my own complaints with them in different areas. Yes, they have made more than one mistake. I believe most coins get overgraded and some get undergraded. There are number of honest dealers that will tell you that a lot of PCGS/NGC coins are overgraded. Some call them "coins with plastic time bombs, ticking away...." That is why I strongly advise people to never buy coins "sight unseen" with no return privileges unless they are buying from a dealer that they trust whom has seen the coin himself. To avoid getting burned by an overgraded coin, just cover the grade and show the coin to a few other dealers. Get their opinion of the grade. Last of all, just use some common sense: if you don't agree with the grade or you don't like the coin, don't buy that coin! With over 6 million certified coins out there to choose from, it's pretty easy to find another one.

Try looking for some of those mistakes that could have (should have?!) been graded higher. They are much harder to find (most get bought up by dealers who break them out of the holders to sell them at a higher grade), but they are out there.

Despite all their faults and failures, these grading services are the best thing out there. On the other hand, if you buy a non–certified (also known as *raw* in coin dealer language) coin or if you buy an investment quality coin certified by someone other than the two services mentioned previously, there is little chance that the grade and/or the value will be as represented. The ultimate analogy is this: No one with little or no mechanical skills would walk onto a used car lot and say "I have $5000, pick me out a car" without getting it checked out by a reliable independent mechanic. The grading services are *your* mechanics.

Find a dealer who has proven his honesty and integrity. One that would never cheat you, even if he did have an opportunity. Dealing with the right person is actually one of the most important criteria for making profits in numismatics. Remember that expertise does not equal trustworthiness. Just because a person is the foremost expert in his field does not necessarily mean that he is honest. And just because someone has been in the coin business "over 20 years" (or any other length of time) does not automatically mean that he is trustworthy either. As with any investment, investing your money with someone that has more interest in "his own" interest will mean guaranteed losses. Even if you are getting accurately graded coins, if your are paying too much, you are being robbed of potential profits. Hence....

The ultimate coin collecting/investing strategy is this:

Find an honest dealer who will sell you coins at prices as close as possible to the true market that *other* dealers would pay you if you had to sell.

A few tips for buying coins

Never buy a rare coin from the same guy who is promoting it! See what other dealers are selling it for. You will be amazed, if not shocked, at how much you can sometimes save. Dealers who are doing promotions of specific coins usually have the highest mark-ups.

On the other hand, the real test is to call up a few larger coin dealers and ask what they would pay for one. Now that will really scare you!

Question: If dealer "A" is promoting a coin for investment at $600 each, and the most anyone is willing to pay you for one is $450, how much does this thing have to go up before you can break even?

Answer: Way too much!

This brings us to another tip to look for to find a more honest dealer. As a general rule (but not always), the more PCGS and NGC coins the dealer has on display, the more likely he is a reputable, some-what honest dealer. However, keep in mind the opposite is not always true. Some of the larger, most honest bullion dealers have very little interest in rare coins. As such, they are unlikely to have any rare coins of any kind in stock in stock, certified nor not. On the other hand, if the dealer is representing himself as dealing in investment quality rare coins, but he keeps very little PCGS or NGC graded coins in stock, than we suggest you find someone else to do business with.

In the previous bullion section, we referred to the term of "mint" condition, the term given to coins before the are released from the bank the first time for general circulation. Even in the new, "mint" condition there are 11 levels or "grades" to account for the different levels of quality. Brand new coins can still get various scuffs, scratches, and nicks as they rub against each other in bank bags. The grading scale for mint condition, also referred to as uncirculated, runs from MS60 to MS70. The "MS" stands for the term "mint state." We are not going to go into any more details about grading here. Collectors spend years trying to master the skill. And entire chapters of coin books are writing about grading coins.

Since so many books are written of the subject of rare coins and buying or selling them, we will not be going into any more details on the subject here except what we have already said. "Just say no."

SECTION V

YOUR CHURCH AND THE CHURCH

CHAPTER 14

YOUR CHURCH
AND Y2K

Your Church

Although a church does not have to make all the same types of plans as a family, there are several issues a church must address in its Y2K preparations:

Five issues here.

1) Debt
2) Assets
3) Physical vulnerability
4) Cash strategies
5) Ministering to needs (this is where the witnessing part will fit in).

1) Debt

Try to get out of it, even if it is painful. This is not a period of time that a church can gamble with all the risks that go along with debt. If a church needs to sell off some assets then sell some off. Is there a piece of land the church has that it can sell to pay off or pay down some debt? Be creative.

2) Assets

They usually come in three kinds. Real Estate such as land or buildings, stocks or bonds, or cash in the bank. Seriously consider selling off additional land the church owns but does not use. In addition to the very realistic potential of substantially dropping land values, the church needs to consider what can be done with the money from the sale. More on that will be covered in a moment. For churches that have stocks or bonds as investment, consider this next point. From a worldly viewpoint, there is a great risk of these investments significantly going down in value. From a Spiritual viewpoint we have this one comment: Mt 6:19, 20 "Lay not up for yourselves treasures upon earth, where moth and rust doth corrupt, and where thieves break through and steal. But lay up for yourselves treasures in heaven, where neither moth nor rust doth corrupt, and where thieves do not break through nor steal."

Y2K might be the ultimate "moth", for both the Church and for individuals.

3) Internal Systems.

Each church needs to closely examine all its internal systems to insure its own Y2K compliance. Shaunti Feldhahn gives these points to consider for churches: "start a high-priority task force to address your church's internal and external Y2K issues. Immediately appoint and empower a staff member (and very probably an elder, lay leader or other interested church member) to address your church's organizational Y2K vulnerabilities (to ensure that your telephones, computers, buildings, bank, etc. will function properly in the year 2000), develop contingency plans, and prepare the Church for ministry during Y2K."[137] Pastors should get Ms. Feldhahn's book, *Y2K:The Millennium Bug*; it has far more information on how churches should prepare for a Y2K ministry.

4) Cash

Within the context of churches, the cash in this case refers to the money in the bank in checking accounts, savings accounts, and various savings bond type things. We want you to look at this from two different angles: liability and opportunity.

First, the liability angle. The people who handle money in the church know how much responsibility there is related to the task. They also know there is a certain liability attached to that responsibility. This is the reason churches tend not to invest in junk bonds and high-risk stocks. But with the Y2K issue looming ahead, what you think of as the safest place for assets, in a regular bank checking account, may actually be at some degree of risk. The greatest risk is for those churches that have more than $100,000 in the same bank. This is that special "all deposits *up to* $100,000 are insured" number. Above that amount, if your bank goes under, technically any amount over $100,000 might possibly "go under" with the bank.

If you are concerned about the stability of your bank, you may want to shift the church's cash reserves elsewhere. Three suggestions.

Our first suggestion is to put a portion into United States Federal Government backed treasury notes or something similar. Find a fund that operates like a money market account, but is all 100% Federally insured. This is actually about the best option. This is not to be confused with normal mutual funds or regular money market. Avoid those.

Our second suggestion will be a little bit hard for some churches to do. Not because of the logistics, but because of the internal politics (which are always harder to overcome than any actual logistics). We believe that a portion of the funds that churches have, after paying off all the debts, should be designated for post-Y2K ministry. Some should be in "real" cash or Federal Reserve Notes-that paper money stuff. The cash could be divided among 5-10 (depending on the size of your church) church leaders, chosen by the deacons, for safekeeping for Y2K induced emergencies. Your church may already have a "benevolence" fund and/or committee in place. We suggest you use these organizations to disburse funds as needed.

Our third and last suggestion is to deposit the rest in a Y2K–compliant bank. Do not place more than $100,000 in any one bank.

If the banks are temporarily closed, or if withdrawals are limited, a church that chose to keep its money in the bank may not have immediate access to its money. And as a result, it would not be able to as effectively use those funds to meet needs. On the other hand, if things like building funds, saving for a rainy day, plans for the fanciest pipe organ in town, or some other project, are more important than meeting needs and being a witness, then this whole section is a moot point.

Let's look at what risks would be taken if a church actually *did* take some of its money out of the bank. Obvious, theft is the greatest risk. It does have to be considered, and steps taken to protect from it. The only other real risk is the loss of potential investment income. Now for some people, the potential loss of a little investment income on the sacred stash is "too great a price to pay. And the concept of having a pile of cash hidden somewhere is absolute paranoid insanity. Never, never, never! You just can't do that!" (You get the idea!)

In reality, if your church did "cash in its chips" for a few months, and if January 2000 comes and goes with little trouble, then it could put all the money back "from whence it came" before it was taken out. In fact, if banks seem to be operating smoothly, then the money could be put back as soon as the second or third week of January. In our opinion (as many of these ideas are-just opinions), if there are major banking problems, they should surface the first week.

5) Ministering to needs

If there are significant problems due to Y2K, there will be significant personal needs as well. Churches with available funds may be in a fantastic position to meet needs and as a result have a fantastic opportunity to witness to many it could never otherwise reach. Reaching the previously proud, self-sufficient, and self righteous who have now been humbled by this and have new found needs could be, as Larry Burkett says, "the greatest opportunity that most of us will ever have in our lifetime to help people and share the message of Jesus Christ." Churches, as a whole, need to embrace the concept of preparedness as a means to minister to many in their community.

While some church leaders are urging their readers and followers to "run and hide," we and other pastors are exhorting Christians to stay and minister to the needs of their community. Patriarch magazine quotes this verse to justify hiding during this crisis: "Come, my people, enter your chambers, and shut your doors behind you; hide yourself, as it were, for a little moment, until the indignation is past. For behold, the Lord comes out of His place to punish the inhabitants of the earth for their iniquity; the earth will also disclose her blood, and will no more cover her slain." Isaiah 26:20-21

Here is why we disagree:

• One could interpret the verse as exhorting us to stay in our own homes. Enter *your* chambers, shut *your* doors. It doesn't say to move somewhere else.

• This conflicts with the entire gospel and the concept of the golden rule. As Pastor Jonny Crist puts it, "And the people that say you'd better leave the big cities and you'd better...get out,' I think, where is the heart of Jesus in that? Why wouldn't Jesus stay right where hurting people live rather than trying to get away from them?" God's word compels us to serve those around us. "Then shall the King say unto them on his right hand, Come, ye blessed of my Father, inherit the kingdom prepared for you from the foundation of the world: For I was an hungered, and ye gave me meat: I was thirsty, and ye gave me drink: I was a stranger, and ye took me in: Naked, and ye clothed me: I was sick, and ye visited me: I was in prison, and ye came unto me. Then shall the righteous answer him, saying, Lord, when saw we thee hungry, and fed thee? Or thirsty, and gave thee drink? When saw we thee a stranger, and took thee in? Or naked, and clothed thee? Or when saw we thee sick, or in prison, and came unto thee? And the King shall answer and say unto them, Verily I say unto you, Inasmuch as ye have done it unto one of the least of these my brethren, ye have done it unto me. Matthew 25: 34-40 If we turn our backs on this hurting nation, God should turn his back on us. "What brings this home to me is the thought of an elderly woman all alone in a major US city. If the power goes out and the heat goes off, who will help her? Who will know she's there? If the church doesn't have the responsibility to help this widow, who does?"[138]

• We should love our neighbors more than ourselves. "As Christians, we should be willing to enter into a time of turmoil, enter a time of potential danger in order to love our neighbors. And that's really our model with Jesus Christ. He didn't stay in heaven in His nice cushy situation, He came down here in the dirt with us and sacrificed Himself on our behalf..." says Shaunti Feldhahn.[139] The run and hide philosophy would have us turn our backs on them when they need us the most. "And they *overcame* him by the blood of the Lamb, and by the word of their testimony; and *they loved not their lives unto the death.*" Revelations 12:11

• God has prepared His Church for such a time. "I don't believe that God geared up this army to go run and hide every time we have a problem.... Not only do you want to bring people in to feed them, which helps their body, but you want to give them the message of Jesus Christ, which helps their soul."[140]

• The "run and hide" crowd tells us that "it is a godly work to prepare to survive, so making preparations for survival is the second proper response to Y2K."[141] (The first response they recommend is repentance, which is, of course, a very crucial step in preparation.) However, we feel our earthly survival is secondary to obeying the scriptures in church building. "Life and survival are not the primary goal, but continuation of the Church and Christian Witness."[142]

• Which brings us to our next point, fleeing the cities in wild-eyed terror with our canned goods, bags of wheat and kerosene lamps will ruin our witness. What would be the appearance of your actions to your current neighbors? How would your actions affect your witness to each and every one who knows you? Keep in mind, as Christians, our actions speak far louder to those who know us, than any words we can ever tell them. We will ruin our testimony if others perceive us as "wackos." Let me again quote an apparent non-Christian who has an opinion of all the hysteria surrounding most Christian Y2K publications: "It's been obvious to me for quite some time that the rising noise levels from extremist viewpoints would soon get in the way of moving forward with needed Y2K efforts. Although it may serve the needs of some to decree that Y2K is a sign from God to repent and prepare for the Rapture, I am not of that extremist camp."[143] He is right. We say that extremist survivalist recommendations cloud out the best plan of attack: the best witnessing opportunity of the century!

• If all Bible believing Christians flee the urban areas, there will be no salt left behind. The prediction of rampant rioting and crime may be self-fulfilling. The very presence of a praying, actively ministering church in an area can avert the forces of evil and provide for the needs of the community. That is why our preparations as a church are crucial to successful survival of the gospel message.

Franklin Saunders points out this fact, "Consider the Church in Germany after World War II. At the war's end, every other institution of government came down. There was no other authority for the occupation forces to deal with. The German Church became, quite literally, both government and provider, even parceling out food to the people.... After the time the Church served as Germany's only government, revival and repentance swept the nation."[144] Now we have that chance to serve the church's primary directive; we should not neglect the chance to prepare for it. "Larry Burkett believes an organization already exists that can begin Y2K preparation now in every community. 'We have the best potential organization.... We just don't use it. We don't work together, mostly because we don't need each other too much. It's called the Church. What you and I cannot do individually, we can all do collectively.... We have to learn to work together.'"[145]

• If all the Christians run away, someone else will minister to the lost. Many people already know that as a general rule, Mormon husbands are required to have on hand enough food to feed their families for year. It would be a sad state of affairs if the Mormon church could minister to millions from their abundance, but the true bride of Christ had nothing to share. Even communities and governments are planning to offer help to people in need on New Year's Eve. A community college in our area plans to have its building open and ready so that, "If utilities were to fail, community buildings like ours could become important sites to serve as shelters, meeting places, staging areas for planning and so on. We've already met with [city] officials to share information."[146] The government has already taken God's glory and the Church's work by helping those in need. That is one reason why our nation has turned its back on the Lord and His Church. We need to follow the Lord and minister to the needy ourselves.

So, how do we prepare our church for the greatest evangelistic outreach of our time? Joseph can be our model. "He saved up...during a time of great prosperity, he got other people to save up and prepare. And then when the famine actually hit, he was not only able to protect himself and his country—Egypt, at that time—but to bless all the surrounding communities, as well as saving Israel in the process, of course. And this is an incredible model for how we as Christians can really have an impact on our communities during a potential time of turmoil."[147]

How to get your church prepared

- Schedule evangelism training and classes. If your Church does nothing else to prepare for Y2K, it should at least do this. This is something the Church needs to be doing anyway.

- Stock up the church pantries with preparedness supplies similar to your own supplies: food, flashlights, oil heaters, kerosene lamps, medicine, tools, blankets, first aid supplies and more. In fact, these supplies should be on hand at all times to provide for unexpected needs. Larry Burkett makes this very important point: "What we can't do individually, we can do collectively. What we can do within the churches is far more than what we can do individually. We can store more food, we can store more water, and we can buy a generator for the church. And you can have a place to sleep if your power's off. Because in the Northern Hemisphere, if this happens, it's in the middle of winter. If you have a wood stove and extra wood outside, it won't be a problem, but if you live in a small apartment somewhere, having no heat could really be a problem. So you've got to have an alternative. Our alternative is collective. We need to do it together, as a body of Christ, within the local church."[148] See Chapter 8 about why preparedness is a good thing.

- Schedule prayer meetings for repentance and prayer for your community. Larry Burkett recommends asking "God to transform this Y2K problem into an opportunity for the Church."[149]

- Check out these web sites for church and community Y2K preparation and ideas.

Joseph Project 2000: This new non–profit organization is dedicated to helping Christians and communities prepare for Y2K. Their primary focus is on how to use Y2K as an opportunity to minister to believers and unbelievers and bring glory to God. We agree with their motives and highly recommend contacting them for

excellent information on how to organize Y2K preparation and plans.

Joseph Project 2000
6406 Bells Ferry Road
Woodstock, GA 30189
Phone: 678-445-5512 Fax: 678-445-5503
E-mail: info@josephproject2000.org
Web site: www.josephproject2000.org

The Cassandra Project: a grassroots group trying to
organize communities to face Y2K problems.

Web site: www.millennia-bcs.com

Of course if Y2K develops significant problems, there would be
problems within the Church's own body as well. And then there are
the missionaries, who may find that nearly all sources of funds virtually cut off. If a Church has access to real money, when very few others have access, they can be an incredible blessing to these desperate
missionaries along with many others.

An appeal to all Church leaders: Pray, seek counsel, and bring this
issue up with your congregations. And as you do, remember this as
well, just whose money is this anyway? Yours? The Congregation's?
Or could it possibly belong to God? In the words of the religious
catch–phrase that is running the risk of becoming so worn out as to
be almost useless, "What *would* Jesus do?" As we see it, if you compare between the loss of a little interest as opposed to the potential
risk of not being able to serve the needs of our friends, families and
neighbors, it makes the call to action clear.

Looking at the downside of the stocking up strategy, what would
happen to all this wasted money spent on a bunch of food and such,
if there are no major problems due to Y2K? Just maybe there would
be some families in your Church of lower income who could actually use some of this food and other necessities. What a radical thought!
Use the Churches financial abundance to give to those in need! (Can't
say it was our original idea, though. Someone else came up with it
first We "stole" the concept from The Book of Acts!) Larry Burkett
points out, "Granted, you may be ridiculed if you cleared out two

rooms of your church to store food in and then you didn't need the food. But, you know what, there are always hungry people, you can get rid of that food, no problem."[150]

If the families of your Church are so blessed that none of them have any personal need for this unfortunate, useless Y2K stockpile, chances are there just might be someone in your community who is in great need of such goods, even in the best of economies and the best of times. And you might even have a chance to witness to them!

Wasn't that what it's all about? Train the body of the Church to go out to the world.

CHAPTER 15

IS YOUR CHURCH THE RIGHT CHURCH?

(Please note that we titled this chapter "Your Church" as in the Church that you have chosen to attend each Sunday. NOT as in "The Church" as referring to the total body of believers who make up The Church as a whole, such as referred to in the Bible. The relevance of the distinction between the two will become self-evident as you read on.

We believe, if you move to the country, the most important question is this: Where would you attend Church?

We believe the issue of what Church a person attends is actually of greater importance than our response to the whole Y2K issue. And that is why we feel extremely compelled to do significant coverage of this issue.

For our family, this issue and this issue alone is the single main reason why we live where we live. We have a fantastic Church and unless we felt led of the Lord, we would not move anywhere outside of this area unless we found an equal or better Church. And while we are on this topic, we are going to take some time to preach at you. But first, some background.

At one point in our lives, we seriously examined the issue of where we should live. We were self-employed with the mail order business The Home Computer Market. We worked out of the house and had no store or office to tie us to any specific location. So we literally had the freedom to live wherever we desired to live and still be able run our business. We felt a desire to move, but we weren't absolutely sure. We examined all the options: in the mountains, by the ocean, totally

269

away from snow (that got the most votes!), centrally located geo-graphically, closer to which relatives, and such. We even did some house shopping over 1500 miles away from where we were current-ly living. To make a long story short, here was our conclusion. We ended up staying put.

Why? To best answer that question, we need to look closer at the decision process. (Now, if you think this is all getting too drawn out, and has nothing to do with Y2K, just keep reading and hear us out!)

Since we were not specifically locked into one location due to a job, we needed to establish criteria for why we would choose one location over another. Yes, beauty of God's creation around us was important. So was having no snow. (It wasn't the snow we hate; it's the cold that comes with it!) One set of Grandparents lived in Minnesota; the other set lived in Florida.

As we looked at all the criteria, we did not really see anything that made any specific location stand out above the rest. That is until we started house shopping in what we were thinking might be our final choice for a new location. After a week of looking at about a dozen houses, we tried to find a potential Church. We started by asking a few people we knew in the area for some recommended Churches. We also went down the list of Churches in the yellow pages. We had a list of a few basic questions about specific Christian doctrines, and we asked what type of music they had in the services.

In this community of approximately 15,000 we found only two Churches we felt worth checking out. We left the first Church during the intermission between Sunday School and the main Church ser-vice. Based on the two songs we heard during the intermission, we heard enough. We would not allow that kind of music in our own house, so we certainly would not want it in our House of Worship.

The music in the second Church was fine. The preaching was sound. But there were just some things about it that we knew we would not fit in.

But then, the first Sunday we were back in our own home Church, we knew THIS was where God wanted us to be.

The point of all this?

Ignore the Y2K issue for a moment.

What is the number one, single most important thing in your life? We hope you are saying "My salvation; knowing we will spend

270

eternity in heaven with Jesus Christ."

Going from there, we hope your priorities would be a mixture of the following things:

- Your family's salvation
- Your relationship with God
- Your family's relationship with God.
- Your Spiritual growth
- Your family's Spiritual growth

There are a number of commonalties of those top priorities. Obviously, God is the primary one. Another very critical common thread to all these top priorities is your Church.

All other priorities should follow from there, such as my children's education, happiness, finances, home, job, etc.

Unfortunately, based on the many hundreds of Christian families we have spoken to about their own families' Churches, most of them do not seem to make the link between the importance of these top spiritual priorities and the role their Church has in their development. We have spoken to hundreds and hundreds of Christians who are either dissatisfied with their Church or thinking of leaving their Church. But very few of these families have expressed the willingness to actually pull up stakes from that Church and make the painful sacrifices necessary to attend a better Church which is further away or perhaps 'less to their families choosing' for one reason or another.

"But this is where our friends are."

"The pastor is so energetic!"

"The messages are so uplifting."

"We hope to change it."

"But this is where we have always gone."

"We know it may not be the best Church, but our children like it so much here."

"It's so close."

271

"We love the music"

"(Fill in the blank)"

We are not even going to waste the space picking apart the error of this kind of thinking. We hope you can see it for yourself.

"We have a witness and/or a ministry within the Church. We are trying to turn things around." This is the only argument that even comes close to holding water with us. But still, we think that one is wrong too. Your responsibility to your own family's spiritual growth far exceeds your responsibility to your Church's growth. Especially for a Church that is heading in any direction than the one it should be heading.

We were confronted with, and forced to deal with this issue personally.

We were attending what we thought was a wonderful Church. It was a small, young Church with a core of extremely committed believers. These were people who were willing to sacrifice their personal preferences and desires in order to live a Christ-centered life, committed to the ways of the Bible. These were Christians we could look up to as good examples to follow and to challenge us to more Godly, Christ-like living. Not driven out of legalism, but out of a love for God. Family inclusive worship was standard, homeschooling and homebirth was the thing to do, and children were, by all appearances, considered an unmitigated blessing. (The pastor had 7 kids and one on the way. That same man now lives in Colorado and has 10 children. In his words, he says with a smile, "It doesn't count until you get to double digits!") Along with our usual services, we would get together for an extended prayer meeting about once a month. These prayer meetings usually lasted 3 to 4 hours long. Even more amazing was that most of the congregation would attend most of these lengthy sessions. We did street preaching. We went door to door on evangelism campaigns and to invite people to our Church.

We described this Church to others as the closest thing we had ever seen to what the New Testament Church in the Book of Acts was like.

And then things started to change.

The old leaders left to start other Churches, and new leadership was bought in. These new leaders seemed to be less concerned about the spiritual growth of the Church and more concerned for the body count growth of the Church.

In short, instead of bringing the Church to the world like they had been doing, instead of trying to equip the members for spiritual maturity and to go out to witness to the world, they brought the world into the Church. And they seemed to kick the Bible out the door in the process-literally! They asked us to leave our Bibles at home! The leadership did not want us to be thought of as a bunch of 'Bible Thumpers' and they did not want to offend or turn off first time attendees or "seekers" as the new people were called.

Sunday morning services were called "seeker" services and were designed to entertain, captivate, and relax the audience. An alternative Friday evening service was started so you could have your Church and not ruin your weekend too. God forbid, we should wreck all these newcomers' Sabbath! (We know, the Sabbath is actually on Saturday. You know what we mean!)

The music was changed from very Godly and worshipful to a heavy jazz and rock influence. Gone were the simple choruses and beautiful scripture songs we used to sing. Instead we had a "professional" style rock-n-roll band complete with electric guitars, synthesizers, electric drums, and even a set of bongo drums! Not just anybody could sing, teach, or share; now you had to be rehearsed, polished, approved and "winsome." This was designed to be appealing and attractive to the new attendees. The Church leaders and members of the Church band were consciously far more concerned with the entertainment value of the music than they were concerned with the spiritual value. The general tone of the music was specifically geared to (in our words, not theirs) stimulate hand clapping and a desire of the listener to sway or dance with the beat. Now, we have nothing against bright, uplifting, joyful music. But when the effect of the music is to appeal to the flesh, far more then it appeals to the spirit, then the music is carnal and fleshly.

Drama skits were added to liven up the time spent "at church." Many of them were actually pretty good. But others were just awful in their message!

The sermons were designed to be to be energizing and uplifting. "Messages were designed to take common difficulties that people face and apply biblical solutions to them, and hopefully lead them to a relationship with Christ." (Actual quote from church newsletter!!!) As further part of this new philosophy, sermons only briefly referenced any scriptures. Often the message was a common problem we all face and how the speaker solved his problem, through his own personal insights and experiences of how God worked in his life. Or as we often sensed what was unspokenly being taught, "Listen to me, I found the answer. Learn from what I did. It will also help you. See what Christ can do for you too!" A Bible reference is thrown in here and there but no real serious Bible study took place.

Error is inevitable when we conform God's Word to fit our needs. Contrast that Church's approach with the way Mary Pride puts it here, "Instead of reformulating biblical doctrines to make them relevant to modern man, let us strive to have Scripture mold our thought forms, so that we may be transformed into His image from glory to glory just as by the Spirit of the Lord."

Jesus Christ was not presented as the Savior from an eternity in hell for all unrepentant sinners. In fact, the words 'sin' and 'repent' were virtually completely removed from the vocabulary of all the messages.

More about that in a moment.

We could not believe it when we sat through a message that basically said, "God doesn't notice and there is no need for Christians to ask forgiveness once they have been saved."

Probably the greatest theological flaw of this new direction was the approach being then toward salvation.

Their church motto is "Making Christ Attractive to the World" In our Bible, Jesus says that's impossible to do. A friend of ours who left this Church a year after we did had an interesting conversation with his neighbor. Seems this neighbor had visited a Unitarian Church as well as our old Church and concluded that the Unitarian Church was sort of like our old Church. When the Sunday messages are that ambiguous that no clear biblical theology or standard is presented, people will be misled into believing no real gospel or a partial gospel. Both are unlikely to land you in Heaven, but in hell instead!

The ultimate example of this situation occurred to us personally. We will get to that in a few moments.

"Any preaching that relies on story telling or clever outlines more than the Scriptures and the cross is bereft of life changing power and is a flawed and fraudulent gospel that cannot possibly reconcile Holy God and sinful man." (We wrote this quote down along time ago but we forgot who said it) A verse that shined its illumination upon our heart in all this is II Corinthians 4:1-5: "Therefore, since through God's mercy we have this ministry, we do not lose heart. Rather we denounce secret and shameful ways; we do not use deception, nor do we distort the word of God. On the contrary, by setting forth the truth [God's Word-the only absolute truth!] plainly we commend ourselves to every man's conscience in the sight of God. And if our gospel is veiled, it is veiled to those who are perishing. The god of this age has blinded the minds of unbelievers, so they cannot see the gospel of the glory of Christ, who is the image of God. For we do not preach ourselves but Jesus Christ as Lord, and ourselves as your servants for Jesus' sake."

I Corinthians 14:23; "But if an unbeliever...comes in...he will be convinced by all that he is a sinner and will be judged by all, and the secrets of his heart will be laid bare. So he will fall down and worship God, exclaiming, "God is really among you!" This is the kind of church Christ wants-not a church that could be mistaken for the local new age center (Unitarian)! Colossians 1:28 "We proclaim him, admonishing and teaching everyone with all wisdom, so that we may present everyone perfect in Christ." This Bible teaching is totally inconsistent with the goal of making sure, as they put it, "messages were designed to take common difficulties that people face and apply Biblical solutions to them, and hopefully lead them to a relationship with Christ." This is not the GOSPEL!

Aside from the services themselves, these last two examples will give you an idea of the extreme to which the leaders were taking their new approach.

A homeschooling mom wanted to include an announcement for a homeschool interest meeting in the Sunday morning program. At first the secretary indicated it would be no problem. A bit later, the secretary called back and said the Pastor said no go because "We are a seeker church." and "We don't want to offend anyone." All in all, this Church's prevailing message is one of tolerance, relativism, and lukewarmness. They eliminated any hint of conviction, Godly standards, and the principle of reaping what you sow.

Despite the dark picture we have painted here, this Church was not a Cult. They do believe in the Bible and Jesus Christ as God come in the flesh for the salvation of the world. And they had many wonderful things going for it. We dearly miss the deep, deep fellowship for our 'small group' which we were a part of there. But in their zeal to increase membership, they made too many compromises that we could not live with.

We were personally challenged to stay in this Church, to be a witness to others in the Church. But it was a private meeting with the senior Pastor that confirmed our need to leave. We discussed the music. We told him the Bible tells us the function and purpose for music in the Church is for worship and praise, not entertainment. The pastor was firm and committed to the direction they were taking with the music.

The leaders were convinced they had found the right formula for Church growth and they were committed to sticking with it. He pointed out the amazing growth as evidence that what they were doing was right. We disagree. Success does not equate to meaning you are doing God's will. Communist China is the largest country in the world and growing. That does not make it right. The Jehovah Witnesses and Mormon churches are growing fast, but that does not make them right. Doing things God's way, as guided by his word in the Bible is what makes things right.

Most importantly to this meeting, we expressed to the senior Pastor that we were feeling spiritually starved, that the emphasis of the Church was to bring new people in the door and only offer watered–down spiritual milk, with very little real meat. He confirmed our perspective as being basically accurate. He agreed with us. And we all agreed that if our family desired greater spiritual meat, we needed to go somewhere else to find it.

And we did.

It was a very sad, traumatic experience. Nearly every friend we had and most the people we knew attended this Church. We had centered our lives around this Church, and now we were compelled to leave everything behind. We kept in touch some with a couple of families. But for the most part, we had to choose a different path.

We have a couple sad footnotes to this story. These are the ultimate examples of what we were referring to previously regarding the spiritual danger of how this former Church of ours was approaching the

Gospel.

Our next door neighbor had been having marital problems while at the same time seeming to have a fairly new spiritual interest. We had been doing some informal counseling with him, as a neighbor and as a friend. One evening, we presented the Gospel to him in complete detail. He had never heard of this before. It was all new to him. He was not sure about accepting the message personally, so he rejected it. Someone's rejection of the Gospel would not be particularly surprising except, much to our shock, we found out he had been attending our former Church for 6 months at this point!

The couple got divorced, sold the house and moved away.

What makes this story even sadder was what we found out about a year after we had presented the gospel to him. We saw him at a local park and started talking. He told us that he was now in charge of the single adults group at that same Church! We never discussed his salvation status, but we can only hope and pray that the Holy Spirit cultivated and germinated that Gospel seed that the Lord had planted in him a year earlier.

Why are we going into such detail about the flaws of this Church and what does this have to do with Y2K? Nothing and everything. If the spiritual growth of your family is a not issue, or of lesser importance to you, then all this about Churches has nothing to do with Y2K.

On the other hand, if you do place spiritual matters above all else in importance, then this has everything to do with Y2K. If this is the case for you, then every decision you make, especially the ones as major and important as those relating to the Y2K issues, needs to be based on the spiritual ramifications that those decisions will have in your life.

BOTTOM LINE

The bottom line is this: Before you try and put your "Y2K" house in order, you need to start getting your spiritual house in order, which is more important than Y2K. And which church you attend has a major impact on your spiritual house. Or putting it another way, you need to make a decision that the spiritual growth of you and your family is the number one thing in your life which all other decisions are based around, and take a 'backseat' to. And getting your family involved with the best Church possible is critical to that goal. Y2K or not Y2K.

CHAPTER 16

CHOOSING
THE RIGHT CHURCH
FOR 2000 AND BEYOND

By now, some of you have come to the conclusion that you need to find a better church. Of course others are certain they are in the church that God wants them to be attending. And then there is a third group of you who are just not sure if you should leave or stay.

This section is for those of you in the first and third groups.

There are the fleshly reasons for attending a church, and then there are the Biblical reasons for attending a church. We discussed, earlier, some of the examples of fleshly reasons for joining a church. What we are going to cover now are Biblical things to look for in a Church. This same material can also be used to evaluate your current Church to see if you should stay or find a more Christ–centered, Biblically–focused Church.

This first list is things that are based on unarguable Biblical truths. If a Church's doctrine does not follow the Bible's truth in any of these areas, you need to strongly reconsider whether or not this Church could be where God wants you to be, especially if they are teaching things which are contrary to God's word.

Absolute Standards

• Verbal Inspiration of Scripture

If a Church does not believe that the Bible is the irrefutable, inspired word of God, there can be no basis whatsoever for true Biblical teaching. This results in Church leaders drawing from their own wisdom and placing wisdom that over the authority of God's word.

• Virgin Birth

To deny the virgin birth of Jesus Christ is essentially to deny the deity of Jesus Christ as God come in the flesh, conceived by the Holy Spirit, with God as the Father.

• Hell

For whatever reason, many Churches deny the existence of Hell. Maybe because they are in denial of where the consequences of their sin will cause them to end up.

• Resurrection

Without Jesus Christ's resurrection, there is very little meaning to what Christ did on the cross. Christ's resurrection validates the entirety of our Christian faith. As Paul put it in 1 Corinthians 15:12-22, "We are all fools and believe in vain if Christ did not, in fact, rise from the dead."

As we referenced earlier in this Handbook, Jesus' resurrection is the key factor which separates Him from all other "prophets", religious leaders, and self-alleged deities. Furthermore, Jesus' resurrection is one of the critical pieces of evidence that validates all of His Deity and all His teachings.

• New Birth

1 Corinthians 2:14 But the natural man receiveth not the things of the Spirit of God: for they are foolishness unto him: neither can he know them, because they are spiritually discerned.

• Blood atonement

This brings us back to the issue of salvation. The Bible says that without the shedding of Blood, there can be no forgiveness of sins. There is no source of forgiveness, no way to heaven, except through Christ's shed blood.

• Salvation through grace alone

Salvation is either based on what we do and don't do. Or it is based on the blood atonement Jesus Christ made for us on the cross, on our behalf, to forgive us of our sins. Either we are trying to earn our salvation through our own actions and efforts, or it is the free gift of god, as the Bible puts it.

• **Heaven**

It is hard imagine a religion that does not believe in the concept of Heaven, but some do not.

What we have just covered are Biblical truths that are not subject to personal "interpretations" of differences in doctrine.

• **Meat sacrificed to idols**

As given to us in the illustration of meat offered to idols, there will always be things which very Godly, mature Christians disagree on. But the items just covered above are not subject to such disagreements. A Church teaching anything contrary to these basic, fundamental principles is teaching a different religion - not Christianity. Or they are attempting to teach Christianity, but in error.

What we are going to cover in this next section are things which are important to find in a good Church, even though they are not items of major Christian doctrine.

Which "Gospel" is preached?

The issue of "Which Gospel" a Church preaches goes beyond the impact of just salvation. Without the Holy Spirit within us, God's Word makes very little sense, and is even confusing to the lost. Without the guidance the Holy Spirit gives us, it is nearly impossible to understand, let alone live by, many of the principles and truths of God's Word. If a Church preaches any other message other than the clear, true Gospel, then a potentially dangerous and sad situation results. Lost souls will be sitting there in the pews, Sunday after Sunday, without realizing their desperate situation and plight. If they are not hearing the true Gospel, how can they get saved?

What is even more dangerous about this situation is that since they think they are already on their way to Heaven, there is no motivation to seek after true salvation! ("Blissfully ignorant" as the saying goes.) In fact the problem is even worse, because they think they are already saved, they will be all the more resistant to the message of the true gospel.

We are not against bringing the lost to Church, and some do get saved in Church. But this is a different issue from having the lost, who think they are saved, as members. One of the areas where this issue really hits home is the potential impact on your own home.

One of the primary functions of the Church is to re-enforce scriptural convictions amongst the families. It is very hard, and sometimes impossible for people without the Holy Spirit inside them, to guide them to live by Biblical convictions. Even worse is the "socialization" pressure on our own children by the children of families who do not live their lives based around Biblical convictions, no matter how inconvenient and no matter what the price. Every parent has heard the equivalent of "But the Jones's kids get to..." Or your children have picked up some bad habit, attitude, or influence from some friends without the same standards.

Fellow Believers, we need to be surrounding ourselves with Godly families with the hope that the influences of these more Godly Christians will rub off on our children or us! But sadly, you will not get that in a Church that does not preach the true Gospel of repenting from sin, and accepting Christ's shed blood for the atonement of sin.

Bringing the world into the Church

Is your Church trying to win the world by trying to mimic the world? How can you bring people to the light, by using darkness? This is the very slippery slope many churches are descending. Unfortunately, due to being blinded by simply looking at Church attendance numbers as the most important factor, these Churches are looking to a few "mega-churches" for examples, rather than turning to the Bible. If we search the Scriptures, we will see that it is not the job or purpose of the "The Church" to win souls. That is the job of the members of the body. The responsibility of "The Church" is for the equipping of the saints to do the job of going out to save souls, and to equip them to maturity in Christ.

Music

Again, the key here is to look to the Bible to see what it has to say about the purpose and function of music.

If the music in your Church is more designed to appeal to and stimulate the flesh and to be "uplifting", than it is missing its true purpose.

It is very difficult for us to keep this section short because of the vast importance of this critical issue.

If there is compromise in the music, than there will be compromise in other areas as well.

If the music is designed for and centered around being interesting and appealing to the newcomers, than it going to be wrong and is NOT attractive or appealing to God. We can say this factually because (if you adhere to Biblical truth) the Bible says that God's ways are NOT man's ways. What appeals to the carnal man is the opposite of what is right and what is God's way. On the other hand, Satan's agenda is centered around appealing to the things which appeal to Man's flesh and pleasure and self–gratification. The Lord desires us to worship him in spirit not in the flesh. John 4:23-24 "But the hour cometh, and now is, when the true worshippers shall worship the Father in spirit and in truth: for the Father seeketh such to worship him. God is a Spirit: and they that worship him must worship him in spirit and in truth." If some claim that this type of music wins the lost, keep in mind that God's Word specifies preaching as the way lost people will be saved. I Corinthians 1:21 "For after that in the wisdom of God the world by wisdom knew not God, it pleased God by the *foolishness of preaching* to save them that believe."

We will give you an example of this in our own lives. One of our family's favorite hymns is one that none of us liked when we first heard it. The first time we heard it, we didn't like it from the very first line of the very first verse. Basically, it did not appeal to our personal musical "taste" and preferences. And it was a slow, boring song.

About the fourth time we sang this song in Church, our Pastor told us to sing it as a prayer to God. As we did that, the song took on a whole new meaning. As a prayer, the song had such sweet beauty and reverence. Instead of trying to enjoy the song in our flesh, the song was now ministering to our soul. It is now a song we sing to our children at bedtime!

You need to make a decision as to what you want out of your music. Do you want it to entertain the flesh or minister to your soul? The very nature of the two goals tends to be in direct opposition to each other, so it really comes down to choosing between the two. You can't have it both ways. And, what does God intend the purpose of music to be?

Standards

Do you want to attend a Church that is full of people who choose their standards based on what makes sense to them and is convenient to their chosen lifestyles? Or do you want to attend a Church where people purpose to live their lives based on Biblical convictions, no matter what the price or how "inconvenient" it may be. And, of course, where the Church teaches convictions and living by convictions. Not because you have to, but because it is the right thing to do, if you are living a Christ centered life.

An interesting side note to this. It has been our experience that where the teaching focuses on living your life with Christ as the center of your life, there is not much of a need to teach on a lot of specific standards and convictions. These things just tend to work their way into your life as you open yourself to the promptings and leadings of the Holy Spirit living within you. That is how it tends to work in our current home Church. We will be giving some example of that in a few pages.

Check out the true fruit

The best way to understand this is to read the next section where we discuss the true fruit of our current Church and our explanation as to why we attend this Church. Three key visible components to an effective church are these three scriptures:

- Unbelievers will say, "Surely, God is among you." I Corinthians 14:23; "But if an unbeliever...comes in...he will be convinced by all that he is a sinner and will be judged by all, and the secrets of his heart will be laid bare. So he will fall down and worship God, exclaiming, "God is really among you!"

- The members of the church truly love each other. John 13:34-35 "A new commandment I give unto you, That ye love one another; as I have loved you, that ye also love one another. By this shall all men know that ye are my disciples, if ye have love one to another."

- The members do not love their life even unto death.

Revelations 12:11 "And they overcame him by the blood of the Lamb, and by the word of their testimony; and they loved not their lives unto the death."

• The members love God's Word. One example of this is shown by how many members bring Bibles to church.

Our Current Church

Let us tell you about our "thoughts and feelings" the first day at the Church we attend now.

It was too small. We did not like the music. (Too many "slow" hymns.) We thought Sunday School was boring. They only had two Sunday School classes. Adults and children. Nothing else. All the children were placed in one big class. And we thought the main service was even more boring. In a sentence, we were experiencing Spiritual Culture Shock from our previous Church experience.

With all this, you have to be asking yourself "Why in the world did they ever go back even a second time?!?"

Answer: The children.

As we were 'shopping' around for a Church, one of the acid tests for us was looking at the children. As we looked at the children, we were not so much looking to see what potential playmates there might be for our own children. That was actually pretty irrelevant to us. We were not trying to choose a Church based on the entertainment value for our own children. So why look at the children if not for potential friends for our own? We looked to see the fruit. The children are probably the best example of the kind of fruit that any given Church is producing. In fact, we believe that when it comes right down to the final analysis, there is hardly a better, faster indicator of the true fruit a Church is producing, and what any given Church is really doing in the lives of its members, *especially the children.*

There were some Churches we attended which we liked the preaching, we liked the music, the doctrine was sound, but the children.... Well, to put it in perspective, we concluded that if we made this our home Church, as our own children got older, we would not feel free to let our children do much associating with the other children in the Church. We saw strong rebelliousness and very worldly influences in the teen age Youth Group. It was not the sort of example we would want our children to emulate. Not the sort of thing you hoped would rub off on your own children.

We admit we could be wrong on this point, but based on the conversations we have had with literally hundreds of other Christian families, if we are wrong, we don't think we are very far from the truth.

So back to the previous question: "Why in the world did they ever go back even a second time?!?"

Answer: The children.

When we walked in that first Sunday, it almost like we entered "The Twilight Zone", but entering it in a good way. What we mean by this is this it seemed like we entered a time warp, hurling us back into the 1940's or even earlier.

What was so appealing to that and what does it have to do with the children? At almost any point in history, except for the last 30 years or so, children were bought up to be respectful of others, respectful of their elders, and respectful of all those in authority. Children were also taught to place their responsibilities and the needs of their family above their own pleasure and entertainment. These attitudes and values tended to carry forward into the later years of their adult lives. And the same attitudes those children were raised to have are the exact same attitudes that we, as Christians, should also strive to achieve today. Contrast these things with most children today. We do not think we need to go into any detail about that. We all know how most of today's children compare to those children of previous generations.

Back to "The Twilight Zone." It was like we had entered another era of time. Despite all the things we did not care for, the one thing we *really* noticed was how different the children were. Different in a good way. We had entered a place where the children were the way you would *want* your children to be! In fact, as we looked around, we said to each other, "If our children turned out to be like most of these children, we would consider ourselves to be a 'success' as a parent.

To digress even a little further, allow us to be specific as to what we saw. We saw children of *all* ages sitting through entire Church services. Sure, every now and then, a few would be brought out because fussing or squirming, but for the most part, they sat still and quiet through the service. We saw children who totally ignored trendy fashions and styles, in favor of a generally modest, conservative look.

But one of the things that impressed us most was watching the actions of the teenagers between services and after Church. Instead of huddling together in "cliques" we saw children interacting in a polite, mature manner, with a healthy mix of ages. Perhaps an even more relevant sign was that we say children that loved their brothers and sisters. Not only did the older children *not* try to avoid their younger siblings, but in fact younger siblings were welcome in any group!

We learned very soon, we had very little fear of our children coming home after Church saying, "Mary's parents let her, why can't I?" or "But all the kids are doing it!"

Something else we discovered after we had been attending for a while was that none of the teenagers and almost none of the single adults dated. Nearly every family held the conviction that dating was wrong, and "courtship" was the proper way for preparing for marriage.

(We tell people around the county about this little jewel of a Church we've found, and one of the most common response we get is "Just how far will we have to drive to get there?")

Don't get us wrong, this is not a perfect Church. It can't be, since we attend it! But there are some factors that contribute to it being one of the best Churches we've ever seen. Probably the most important is the very strong focus of both the individual members and the leadership towards having a heart which seeks after Christ, focusing on Christ, and living a life of Christ living in us and through us in all that we do.

Surprisingly, one of the things a person would expect to find in our church, which is actually quite absent, is a heavy emphasis on "don't do this and don't do that." But those sorts of messages as just not given. When one focuses on Christ and living a Christ centered, Christ filled life; most everything else starts to fall into place. We do not have a bunch of written, or even unwritten, rules on do this and don't do that. There just is not a need or a purpose for it.

The hour long drive

We are not the only family who feel so strongly about the value and "special–ness" of this particular Church. There are about half a dozen or more families that drive between 60 to 90 minutes *each way* to attend this Church every Sunday. One family stays in town all day Sundays, just so they can make the evening service too.

Why all the hype?

What is the purpose to all we say about our Church? Is it to brag how spiritual we are? Is it to brag about the great sacrifice people make to get there? (Certainly not us. By God's grace, we are less then 10 minutes away. It is only about six minutes if we hit the traffic lights right!) Is it to revel in our own personal pride of how great a discovery we made? Or how much we have arrived spiritually?

No. It is for none of these reasons.

But we do have a number of reasons why we are saying all this about our Church, and how truly blessed it is by God. We want those of you out there who are discouraged about your own Church, to know that Churches like this do exist, and you just need to find them.

We are going to very loosely draw an analogy from one of Jesus Christ's parables to help illustrate the relevance of all this to you. In the parable of the Pearl of Great Price, the man who found the Pearl, recognized its true value. He was willing to sell everything to obtain it. Without trying to go into a theological discussion as to *exactly* what that pearl actually represents in the original Bible story, we want to draw this analogy. A good Church is like a pearl of great price. And it is worth all the sacrifices necessary to both find one and attend it on a regular basis.

This is our exhortation to you:

Find your Pearl of Great Price. And if it takes a sacrifice of selling all (selling your home, changing jobs, and even moving to a new community or even a new city) to gain this Pearl, it will be worth it and we sincerely believe God will richly bless you and your family for any sacrifices you must make.

We understand there would be exceptions to this. Maybe you have an elderly parent who needs your special attention. (You can always move that parent with you.) You may have a special witness to individuals where you are at now. There may be special needs as to why you need to be where you are now, and you just cannot move. If this is the case, please, at least be open to considering the idea of being willing to drive further to attend Church, if that is what it takes to attend a good, Godly Church.

Please, for your own sake, do not try to make excuses for why you can't find and attend a better Church if you are not 100% positive that where you are already at is the Church God wants you to be attending. Do whatever you need to do to attend a Church that is focused on Jesus Christ as Lord, Savior, and God come in the flesh. A Church where you can know *this* is where your Lord wants you to be. Not just because it is where you want to be.

This is more important than *any* Y2K preparations you can make.

Certainly, if you feel strongly led to move out to the country (or woods, or mountains) and become self–sufficient, we are not telling you not to. (Sorry about the double negatives there!) What we are telling you to do is weigh the costs and make sure you are studying ALL the ramifications involved with your potential actions. And seek wise counsel, including (but not limited to) those God has placed in authority over you.

The reason for our ranting and raving over "Church Choice"

This whole section has been designed to (hopefully) motivate you to find a better Church for your family, if you are not already in a good, Christ-focused Bible-based Church already. Because, in the spirit of this Handbook being **"The Christian's Y2K Preparedness Handbook,"** as a Christian, you can hardly do better preparing for eternity itself than finding a better Church.

CHAPTER 17

CONCLUSION

We hope this Handbook has been a blessing to you, your family and your church. As the Bible says, we wanted to stimulate you to faith and good works. This Handbook came out of a burden the Lord placed on our hearts to help Christians make common sense plans for Y2K and beyond. Too many people were calling us in near hysteria, wanting to spend their life's savings on gold and silver. Some also wanted to sell their homes, quit their jobs and move their families to a remote area. We knew God had outlined other plans in the Bible for His church.

Now you are armed with practical suggestions and resources to actually make a difference in your community, neighborhood, and church for any crisis, not just Y2K. As Christians, we are called to seek those who are lost and minister to those who are sick. Now is the time to seize those opportunities.

Long ago, when the Bubonic Plague swept though London, most of the nobility and wealthy people fled the city. The Puritans stayed behind and nursed the sick and dying. Prior to the plague, the Puritans were severely persecuted and reviled. As Shaunti Feldhahn points out in her new book, *Y2K, The Millennium Bug,* "These believers poured themselves out on behalf of a society that severely ridiculed and repressed them, sacrificing themselves on the alter of love for Jesus Christ and for their fellow man." After the incredible ministry they poured out on the lost and sick, they gained a new respect and many open doors for the gospel.

Too many times, in our country, the Christian church writes off the lost as hopeless, worthless, and deserving of the fate that befalls

291

them. Despite this attitude, believe it or not, God still loves the lost and wants to save them. And whom does He want to minister to a lost and dying world? Us! The redeemed, believing bride of Jesus Christ, who is the *only* light of the world. Let us be faithful to the end.

Matthew 25:31-46

When the Son of man shall come in his glory, and all the holy angels with him, then shall he sit upon the throne of his glory. And before him shall be gathered all nations: and he shall separate them one from another, as a shepherd divideth his sheep from the goats. And he shall set the sheep on his right hand, but the goats on the left.

Then shall the King say unto them on his right hand, Come, ye blessed of my Father, inherit the kingdom prepared for you from the foundation of the world. For I was an hungred, and ye gave me meat. I was thirsty, and ye gave me drink. I was a stranger, and ye took me in. Naked, and ye clothed me. I was sick, and ye visited me. I was in prison, and ye came unto me.

Then shall the righteous answer him, saying, Lord, when saw we thee an hungred, and fed thee? or thirsty, and gave thee drink? When saw we thee a stranger, and took thee in? or naked, and clothed thee? Or when saw we thee sick, or in prison, and came unto thee?

And the King shall answer and say unto them, Verily I say unto you, Inasmuch as ye have done it unto one of the least of these my brethren, ye have done it unto me.

Then shall he say also unto them on the left hand, Depart from me, ye cursed, into everlasting fire, prepared for the devil and his angels. For I was an hungred, and ye gave me no meat. I was thirsty, and ye gave me no drink. I was a stranger, and ye took me not in. naked, and ye clothed me not. sick, and in prison, and ye visited me not.

Then shall they also answer him, saying, Lord, when saw we thee an hungred, or athirst, or a stranger, or naked, or sick, or in prison, and did not minister unto thee? Then shall he answer them, saying, Verily I say unto you, Inasmuch as ye did it not to one of the least of these, ye did it not to me. And these shall go away into everlasting punishment, but the righteous into life eternal.

One Final Note

Our ultimate goal with this Handbook was to not just prepare you to get through this Y2K event. Our goal was to save the lost and to motivate Christians to live more Godly, Christ filled lives. To do this, we strived to convey two main messages. The first thing was to clearly and correctly communicate the details of how to be forgiven (saved) and go to Heaven. The second thing we wanted to convey practical steps was to get Christians to be more Godly and Christ centered in their lives. The two obvious ways to do this are through reading the Bible and to find a better Church to assist you and your family in the process of being more like what Christ desires us to be like. There is one more thing we would like to recommend that would do far more good than nearly any other than the Bible or a great Church could offer you. We would like to suggest the one other thing that would have the next greatest benefit to your life, as well as your children's lives.

At some time in our lives, each of us has heard or read some message that has motivated us to greater achievement or to make changes to better our lives. Of all those things you have read or heard that initially motivated you, how much of a lasting change had it actually made in your life a year later? So often we hear or read a great message that we want to embrace in our life, but a year later we often can't even remember it! Yes, there are those messages we absorb that have a lasting change in our life, but those are few and far between. And those tend to change only a small portion of one's life. Rarely does one hear a message or read something that has a positive, major impact in changing one's life forever.

Our Search, Others' Searches

About the time we got married, in our search to learn more about Jesus Christ and the Bible, we attended a seminar that was recommended to us by our new pastor and by a Christian we highly respected. The first night we attended, it changed our lives forever. The seminar that we attended is called "Institute in Basic Life Principles." (Also known as "Institute in Basic Youth Conflicts") It teaches seven key principles that affect everyone's life. We have seen more lives changed for the better from this seminar than any other seminar or book that we know of (outside of the Bible itself). And on the subject of the Bible, nothing that we have ever seen even comes close to making the Bible real and more relevant to everyday life than this seminar.

We receive no money for promoting this seminar, and have no interest, financially or otherwise, in its operation. If we can do anything to help you train your children for true success, and to better your own life, it would be to invite you to this seminar. They are usually held about once a year in most major cities. If you've ever attended the seminar before, you can re-attend at anytime for free! And we would encourage you to attend again if you haven't done so recently. There is a lot of new material, and much of the previous material has been updated, refined, and reworked for even better clarity and explanation. Also, at some cities they have added a "Children's Institute" for ages 6-12. Our children love it!

There is also an Advanced Seminar for "Alumni" of the Basic. If you found the Basic Seminar helpful, you'll love this one. For more information about any of the seminars, contact us, your local IBLP office (if you have one), or call the national headquarters at (630) 323-9800. Or check out their web site at www.iblp.org.

FOOTNOTES

[1]Capers Jones, The Year 2000 Software Problem, 1998 p. 226.

[2]David Eddy is president of Y2K Service Corps, Inc., a firm that specializes in missionary marketing efforts for Year 2000 services. Mr Eddy has 15 years of software development experience in banking, insurance, and consulting environments. He is the origin of the term "y2k." We will quote him again later in the Handbook . Check out his articles at www.y2ktimebomb.com

[3] Capers Jones, The Year 2000 Software Problem, 1998 p. 226.

[4] Ibid, p. 95.

[5] The Danger of a Dumb Analogy, By David O'Daniel Eddy January 6, 1999, www.y2ktimebomb.com

[6] Capers Jones, The Year 2000 Software Problem, 1998 p. 21

[7] Ibid

[8] San Francisco Examiner, April 19, 1998, Rebecca Lynn Eisenberg, http://www.examiner.com/skink/skinkApr19.html

[9] Ibid

[10] Ibid

[11]Coming Home y2k Newsletter, December 1, 1998

[12]Capers Jones, The Year 2000 Software Problem, 1998 p. 1.

[13]www.cfcministry.org

[14] www.y2k subcommittee.com, November 23, 1998

[15]www.cfcministry.org

[16] Tip of the week, by Jim Lord, January 1, 1999, http://www.y2ktimebomb.com/Tip/Lord/lord9901.htm

[17]Capers Jones, The Year 2000 Software Problem, 1998 p. 56.

[18]Ibid, p. 14.

[19] Mitch Ratcliffe , December 31, 1998, http://www.msnbc.com/news/227483.asp#BODY

[20]Ed Yourdon, Time Bomb 2000, p. 283

[21]Ibid, p. 285

[22] Software Magazine, The Embedded Legacy, by Rick Whiting, p. 23, 4/15/98.

[23]Control Magazine, September, 1998 p. 75

[24]Ed Yourdon, Time Bomb 2000, p. 286

[25]David Eddy, www.y2ktimebomb.com

[26]Control Magazine, September, 1998 p. 88

[27]Ibid, p. 89

[28]Ibid, p. 81

[29]USA Today, September 19, 1998, p.4D

[30] www.y2ktimebomb.com/Bios/dmbio.htm

[31] Dick Mills, Power Failures in 2000, 6/26/98, www.y2ktimebomb.com/PP/RC/rc9825.htm

[32] Dick Mills, Another Myth- A black grid can't be restarted, 8/14/98, www.y2ktimebomb.com/PP/RC/dm9832.htm

[33] Alan Gartner, Telgyr Systems Inc. INFOWORLD Magazine, September 21, 1998. p. 73 www.infoworld.com

[34] San Francisco Examiner, April 19, 1998, Rebecca Lynn Eisenberg, http://www.examiner.com/skink/skinkApr19.html

[35] Dick Mills, Another Myth- A black grid can't be restarted, 8/14/98, www.y2ktimebomb.com/PP/RC/dm9832.htm

[36] Ibid

[37] Dick Mills, Another Myth- We Must Fix all the Bugs To Have Power, 7/17/98, www.y2ktimebomb.com/PP/RC/rc9828.htm

[38] Dick Mills, Things won't explode, www.y2ktimebomb.com

[39] Steve Alexander, Minneapolis Star and Tribune, November 1, 1998 www.startribune.com.

[40] Infoworld, 10/5/98

[41] Steve Alexander, Minneapolis Star and Tribune, November 1, 1998 www.startribune.com.

[42] Infoworld, 10/5/98

[43] Steve Alexander, Minneapolis Star and Tribune, November 1, 1998 www.startribune.com.

[44] Minneapolis Star and Tribune, January 3, 1999, p. D4

[45] *Another Myth, We Need Computers to Synchronize, By Dick Mills, December 11, 1998,* www.y2ktimebomb.com

[46] www.cfcministry.org

[47] *Howard Belasco* ,www.y2ktimebomb.com/Computech/Issues/hbela9846.htm

[48] Ibid

[49] USA Today June 10, 1998.

[50] Rick Cowles, www.y2ktimebomb.com/

[51] top ten business benefits

[52] www.cfcministry.org

[53] Shaunti Feldhahn, Y2K: The Millennium Bug, 1998, p.71

[54] Ed Yourdon, Time Bomb 2000, 1998, p. 119

[55] Infoworld, 10/12/98

[56] Ibid

[57] USA TODAY June 10, 1998.

[58] www.cfcministry.org

[59] USA Today, 08/20/1998

[60] www.cfcministry.org

61*David Eddy*.www.y2ktimebomb.com

62 Minneapolis Star and Tribune, January 3, 1999, p. D4

63 SEC fines brokerage firms

64Newsweek, June 1997

65Software Magazine, 10/15/98 p. 96.

66OCC backup plans

67Infoworld, May 11, 1998, p. 3

68Capers Jones, The Year 2000 Software Problem, 1998, p. 3

69Minneapolis Star and Tribune, October 25, 1998.

70USA Today, 9/9/98

71USA Today June10, 1998

72Newsweek, June 1997

73 www.zdnet.com/zdy2k/

74Minneapolis Star and Tribune, October 25, 1998.

75PC Week, March 2, 1998

76David Eddy, www.y2ktimebomb.com/Techcorner/DE/de9835.htm
"Aunt Millie's Check," September 2, 1998

77Ed Yourdon, Time Bomb 2000, 1998, p. 250

78Ibid, p. 253.

79Miami-Dade County Y2K Response Planning By Chuck Lanza,
www.y2ktimebomb.com

80Software Magazine, 10/5/98, p. 12

81Grant R. Jeffry, The Millennium Meltdown, 1998 p. 158.

82Ibid, 1998 p. 155

83Ibid, 1998 p. 152

84Capers Jones, The Year 2000 Software Problem, 1998, p. 155

85Dan Barkin, The News and Observer, 11/15/98.

86Capers Jones, The Year 2000 Software Problem, 1998, p. 153

87Minneapolis Star and Tribune, October 25, 1998.

88Infoworld, 5/11/1998, p. 106

89Jesse Feiler and Barbara Butler, Finding and Fixing Your Year 2000 Problem,
1998.

90Michael S. Hyatt, The Millennium Bug, 1998, back cover.

91Shaunti Feldhahn, Y2K: The Millennium Bug, 1998, p.158

92http://www.yardeni.com/y2kbook.html

93Evaluating the Economic Impact of the Year 2000 Problem, Dr. Reynolds
Griffith, Professor of Finance, Department of Economics and Finance
http://titan.sfasu.edu/~f_griffith/sweconw.htm

94Michael J. Mandel, "ZAP! HOW THE YEAR 2000 BUG WILL HURT THE
ECONOMY", Business Week, March 2, 1998, available at http://www.business-
week.com/premium/09/b3567001.htm)

[95]Shaunti Feldhahn, Y2K: The Millennium Bug, 1998, p.75

[96]www.cfcministry.org

[97]Evaluating the Economic Impact of the Year 2000 Problem, Dr. Reynolds Griffith, Professor of Finance, Department of Economics and Finance http://titan.sfasu.edu/~f_griffith/sweconw.htm

[98]Minneapolis Star and Tribune, Sunday October 25, 1998.

[99]Ibid.

[100]Minneapolis Star and Tribune, November 29, 1998

[101]LAPD Gears Up For Y2K Computer Chaos, Fort Worth Star-Telegram, January 07, 1999

[102]Computer Reseller News, October 19, 1998, p. 5

[103]Ibid, p. 5

[104]Wayne Rash, Internet Week, 11/23/98

[105]Information Week, 9/21/1998

[106]www.cfcministry.org

[107]Tip of the Week, by Jim Lord, January 1, 1999, http://www.y2ktimebomb.com/Tip/Lord/lord9901.htm

[108]Information Week, 9/21/1998

[109]Howard Belasco, www.y2ktimebomb.com/CP/Personal/hbela9842.htm, 10/20/98

[110]Rolling thunder: When the bug will strike, 12/31/98, http://www.msnbc.com/news/227483.asp#BODY

[111]Ibid

[112]Ibid

[113]*Y2K: It's Closer Than You Think* By Michael S. Hyatt, October 12, 1998, www.y2ktimebomb.com

[114]Michael Alan Aisenberg www.y2ktimebomb.com

[115]Leland Freeman, VP of Source Recovery Company, in Software Magazine, 10/15/98 p. 22

[116]PC Week, 10/5/98

[117] USA Today, 7/8/98, p. 1A

[118]From Cap Gemini Americahttp://www.usa.capgemini.com/Y2K/, http://www.year2000.com/y2knews.html

[119] "The Ultimate New Year's Eve Party?" Normandale Community College Connection Newsletter, Spring 1999, p. 12

[120] A Community Goal: Self-Sufficiency for 14 Days By Chuck Lanza December 18, 1998, www.y2ktimebomb.com

[121] Rebuttal to Devolutionary Spiral, December 9, 1998 By David O'Daniel Eddy http://www.y2ktimebomb.com/Techcorner/DE/de9849.htm

[122] Making the Best of Basics, by James Talmage Stevens, 1997, p. 16

[123]Men's Manual Volume II, Institute in Basic Life Principals, Chapter 11

[124]USA Today, June 8, 1998

[125]Software Magazine, April 15, 1998, p. 24

[126]Computing Today, January/February 1999, Vol.3, No. 1, Page 12

[127]Christian Financial Concepts Newsletter, Issue 244, May 1998, p.1

[128]Susan Conniry, http://members.home.net/shadow-scout/second.html

[129]Susan Conniry, http://members.home.net/shadow-scout/#fourB

[130]Ibid

[131]Ibid

[132]Cooking With Home Storage, Vicki Tate.

[133] FEMA info at www.isd.net/stobin/fema/emfdwtr.html

[134]The Millennium Bug, by Michael Hyatt, p.

[135] Larry Burkett, quoted in Shaunti Feldhahn's book, Y2K: The Millennium Bug, 1998, p. 172

[136] Ibid, p. 173

[137]www.josephproject2000.org

[138]Chris Mitchell, "Faith, Not Fear: The Church's Response to Y2K" www.cbn.org, Pat Robertson's ministry

[139]www.cfcministry.org

[140] Larry Burkett, www.cfcministry.org

[141]www.babel2000.com

[142]Franklin Saunders, www.the-moneychanger.com

[143] Rebuttal to Devolutionary Spiral, December 9, 1998 By David O'Daniel Eddy http://www.y2ktimebomb.com/Techcorner/DE/de9849.htm

[144]Franklin Saunders, www.the-moneychanger.com

[145]Larry Burkett, www.cfcministry.org

[146]"The Ultimate New Year's Eve Party?" Normandale Community College Connection Newsletter, Spring 1999, p. 12

[147]Shaunti Feldhahn, www.cfcministry.org

[148]www.josephproject2000.org

[149]Christian Financial Concepts Newsletter, Issue 245, June 1998, p. 1

[150]www.josephproject2000.org